This book comes with access to more content online.

Quiz yourself, take sample tests,
and score high on test day!

Register your book or ebook at
www.dummies.com/go/getaccess.

Select your product, and then follow the prompts
to validate your purchase.

You'll receive an email with your PIN and instructions.

ACT® Math Prep

2nd Edition

by Mark Zegarelli

ACT® Math Prep For Dummies®, 2nd Edition

Published by: **John Wiley & Sons, Inc.,** 111 River Street, Hoboken, NJ 07030-5774, www.wiley.com

For general information on our other products and services, please contact our Customer Care Department within the U.S. at 877-762-2974, outside the U.S. at 317-572-3993, or fax 317-572-4002. For technical support, please visit https://hub.wiley.com/community/support/dummies

Wiley publishes in a variety of print and electronic formats and by print-on-demand. Some material included with standard print versions of this book may not be included in e-books or in print-on-demand. If this book refers to media such as a CD or DVD that is not included in the version you purchased, you may download this material at http://booksupport.wiley.com. For more information about Wiley products, visit www.wiley.com.

Library of Congress Control Number: 2024932621

ISBN 978-1-394-24226-9 (pbk); ISBN 978-1-394-24228-3 (ebk); ISBN 978-1-394-24227-6 (ebk)

SKY10070913_032724

Contents at a Glance

Table of Contents

Introduction

More than one-third of all graduating students in the United States — an estimated 36 percent — take the ACT (which, by the way, stands for American College Testing). If you're among this lucky bunch, you may be looking toward the future with a combination of anticipation and dread. You may be anticipating college, with its new experiences and first tastes of freedom, but dreading the hoops you have to jump through to get there. Of course, the ACT is one of these hoops. If you do well on this test, you're propelled to your first-choice college; if you do poorly, maybe not.

You're not alone. And I'm here to help. As the author of *Basic Math and Pre-Algebra For Dummies* (Wiley) and *SAT Math For Dummies* (Wiley), I've already helped thousands of would-be college students get through the arduous testing process and move into the exciting life that awaits them in college.

If you're willing to keep the faith and do the work in front of you, I'm confident that you can be among those who succeed on the ACT and take one step forward into your dreams. Let this book be your guide.

About This Book

Although you certainly want to do well on all four sections of the ACT (as well as the optional writing test), this book focuses exclusively on math. The reason I'm playing favorites is simple: For many students (and possibly you, too), math is the toughest section of the test. Because your composite score on the ACT is based in part on your math performance, you don't want math to drag down an otherwise good score.

The math you need to know to do well on the ACT is basically what's covered in high school: pre-algebra, elementary algebra, intermediate algebra, coordinate geometry, plane geometry, and trigonometry. If you're thinking of college, the good news is that you've probably studied a lot of this material already. The bad news is that you may have forgotten it or never felt entirely comfortable with it in the first place.

This situation is perfectly normal, and most students who aren't math geniuses feel roughly the same as you. So your task is to find a realistic strategy to pull together the stuff you already know — even if you're not currently confident that you know it — and tie in the more advanced topics that may have eluded you. Luckily, you've come to the right place.

Everything in this book is designed to take the small amount of math you may remember and build on it so you can use it to tackle typical questions that appear on the ACT. If you've laid the groundwork in school and you're willing to practice, I'll help get you through the test.

My approach in this book is simple: practice, practice, and more practice. The focus is specifically on the types of questions that appear over and over again on the ACT. Even if math isn't your strong suit, when you become comfortable with this core material, you stand a much better chance of getting the kind of ACT score you want to see. And the best part is that you don't necessarily have to read this book from cover to cover. You can hop and skip around to whatever topics you need to work on most.

Conventions Used in This Book

Here are a few conventions to keep in mind as you make your way through this book:

- Variables (such as x and y) and newly defined terms are in *italics.*
- Keywords in bulleted lists, action parts of numbered steps, and answers in the practice tests are in **bold.**
- Web sites are in `monofont`.
- I alternate the range of the answer choices throughout the book. In one question, you see Choices (A), (B), (C), (D), and (E), and then in the one after it you see Choices (F), (G), (H), (J), (K). Because the ACT itself uses this pattern, I want you to become familiar with it before test day. And, yes, test-designers skip the letter *I* (so I do, too). Why? Probably to avoid confusion because it looks like the number 1.

Foolish Assumptions

I'm going out on a limb here, but if you've bought this book, my first assumption is that you or somebody you know is thinking about taking the ACT. If not, you can certainly use it to improve your knowledge of high school math. And beyond that, it also makes a useful doorstop or something to stick under the leg of a *very* wobbly table.

My second assumption is that you've taken a year of algebra, whether in high school, junior college, or some other place. You don't actually have to feel like you *know* algebra. In fact, the point of this book is to build these very skills. But if you've at least sat through an algebra class, you probably know a lot more than you think you know. Really!

Icons Used in This Book

Throughout this book, I use the following four icons to help you keep track of the different kinds of information. Here's what each icon means:

This icon highlights essential information that you need to know, such as key ideas or formulas. Spending a few extra minutes studying these important points is time well spent. You also can use these icons to skim through a chapter you've already studied. Doing so helps solidify your understanding.

Each tip provides an insightful way to approach a question. You may find it especially helpful as a quick way to cut through a seemingly complicated problem.

This icon is like a flashing red light, drawing your attention to a sticky or subtle point that may trip you up if you're not aware of it. Take an extra moment to slow down and make sure you understand the point being made so it doesn't trip you up on the test.

When you see this icon, you know you're ready to put pencil to paper. This book includes dozens of examples of ACT questions, showing you how to answer them from start to finish. Use these examples to better understand a specific type of problem and then refer to them when answering the practice questions.

Beyond the Book

In addition to the book content, you can find valuable free material online. We provide you with a Cheat Sheet that serves as a quick checklist, including the basic supplies you need to draw, where to find inspiration, how to identify common drawing styles, and more. Check out this book's online Cheat Sheet by searching `www.dummies.com` for **ACT Math Prep for Dummies Cheat Sheet.**

Where to Go from Here

I've written this book as a complete reference to ACT math. You can start anywhere you like, dip in, clarify your understanding, and then hop back out with ease.

If you're completely unfamiliar with the ACT, I recommend that you begin with Chapter 1 before branching out. In that chapter, I outline the basics of the math portion of the ACT and provide an overview of the topics that are covered on the test.

If you're ready to get studying, flip to whatever chapter discusses the topic you need help with most. Chapters 4 through 13 provide a thorough review of the specific math skills that the ACT covers. You can focus on these chapters until you feel ready to take one of the practice tests in Part 5 or online. Alternatively, you can take a practice test first to find out which types of questions you can answer easily and which hang you up. To access the online tests, visit `www.dummies.com/go/getaccess`. Select your product, and then follow the prompts to validate your purchase. You'll receive an email with your PIN and instructions.

1 Getting into the Act: An Overview of ACT Math Basics

IN THIS PART . . .

Seeing an overview of what ACT math includes and excludes.

Learning some important test-taking and calculator skills.

Working with the basic four operations, negative numbers, order of operations (PEMDAS), and fractions, plus using formulas to solve problems and facing down word problems.

IN THIS CHAPTER

» Beginning with an overview of ACT math

» Identifying four important steps to success

» Studying the six math topics tested on the ACT

» Figuring out what's not tested

Chapter 1 Reviewing ACT Math Basics

All across the United States — and especially in the Midwest, South, and Rocky Mountain states — high school juniors and seniors prepare for one of the most action-packed Saturdays of their lives. No, I don't mean the prom, graduation day, or a really excellent date with someone their parents don't know about and wouldn't approve of if they did. No, they're preparing for something even more fun than that: the ACT.

Well, okay, back on Earth, it turns out that at least *some* of this is true: It looks like you *will* be taking the ACT (which I figured out the moment you bought this book). In this chapter, I begin the work of helping you get ready for the most fun part of all: the 60-minute math test.

I start this chapter with a look at what's on the math portion of the ACT. Then I provide a bit of advice about four general ways to improve your score. Finally, I list the six math topics that the ACT tests you on, including a breakdown of the specific skills within each topic that you should focus on to do well.

The chapter ends with a sigh of relief as you discover the math that you don't need to worry about on the ACT. (However, the teacher of your current math class may still want you to know something about it, so don't flush it from your memory just yet!)

Getting an Overview of ACT Math

The ACT contains five separate tests, which are always presented in this order:

1. English
2. Mathematics
3. Reading
4. Science
5. Writing (optional)

This book focuses exclusively on the second test, the ACT mathematics test. This math portion lasts for 60 fun-filled minutes and contains 60 questions. Generally speaking, questions appear roughly in order of difficulty.

The test is scored on a scale of 1 (lowest) to 36 (highest). As a rule of thumb, here's how I think of ACT scores:

- A score of 20 is about average, just about equivalent to a 500 math score on the SAT.
- A score of 25 is a good, college-ready score, similar to a 600 math score on the SAT.
- A score of 30 (or above!) is a great score, on par with a 700 math score on the SAT.

TIP

On the ACT, test graders don't take off points for wrong answers — that is, you won't be penalized for guessing. So keep an eye on the time and, just before your time is up, fill in answers to all 60 questions. Make as many educated guesses as you can, of course. But even wild guessing won't hurt your score, so fill in those answers before time's up!

Taking Four Key Steps to ACT Math Success

I like to give the following four important pieces of advice to those students who want to improve their ACT math scores:

- Sharpen your basic math skills.
- Get comfortable using your calculator.
- Solidify your ACT-math-specific skills.
- Take practice tests under timed conditions.

In this section, I give you an overview of these four steps. The rest of the book is devoted, one way or another, to working on them.

Sharpening your basic math skills

By the basics, I mean the math calculations that precede pre-algebra. For example:

- Multiplication tables up to 9×9
- Adding, subtracting, multiplying, and dividing fractions
- Converting percents to decimals and vice versa
- Converting common percents, such as 10%, 20%, 25%, 50%, and 75%, to fractions and vice versa
- Working with negative numbers
- Knowing the order of operations (PEMDAS: parentheses, exponents, multiplication and division, addition and subtraction)

When I say *know this stuff,* I mean that you should know it stone cold — the way you know your own name. If you're wasting precious time trying to remember 7×8 or calculate $-3 - 5$, you'll benefit greatly from investing just a few hours to work with a set of flash cards designed to help you strengthen these skills. In Chapter 3, I cover these topics to refresh your memory.

Getting comfortable with your calculator

You may already be well aware (and grateful!) that calculators are allowed on the ACT. Moreover, because virtually everyone will be using one, you should consider a calculator not just optional, but mandatory.

WARNING

First off, if you're like me, you probably use your phone as your go-to calculator. Additionally, you may use an online graphing calculator (my favorite one is at desmos.com) for help while working on your homework. Unfortunately, on the ACT, you can't use your phone or any other device that gives you online access, so you need either a scientific calculator or a graphing calculator.

If you already own a good calculator and are reasonably adept at using it, just use this quick rule of thumb to make sure it's up to speed: Check to see whether it can do trigonometric calculations like *sin x*. If it can, it's probably just fine; if not, you need to think about an upgrade well before the ACT so you have time to practice with it.

In Chapter 2, I give you some specifics about your calculator. And for those of you who are considering or who already own a graphing calculator, I recommend *TI-83 Plus Graphing Calculator For Dummies* (Wiley) and *TI-89 Graphing Calculator For Dummies* (Wiley), which are both written by C. C. Edwards.

Solidifying your ACT-specific math skills

Studying for the ACT should be mostly a review of skills that are covered in your math classes. However, no matter how hard you work in your classes, the material you covered two or three years ago may not be fresh in your mind. So focused preparation for the ACT can really pay off.

Parts 2, 3, and 4 (Chapters 4 through 13) provide a detailed review of the math that shows up most on the ACT. You can work through dozens of example problems and answer 108 practice questions specifically related to those topics.

Taking practice tests under timed conditions

Time is money, and money changes everything. So it's not surprising that time changes everything — especially on the ACT, where you have only 60 minutes to answer 60 math questions. Purposefully working under low-stakes time pressure adds a useful dimension to your study, especially as you get closer to your test date when the stakes will be higher.

Part 5 of this book — Chapters 14 through 17 — contains two complete practice tests (and answers!) for you to try out, plus access to a third test online. I recommend that you take them under real test conditions. In other words, take them in one hour with the calculator you plan to use on the test and no additional help. You may start out taking the first test as a benchmark before you begin working on the rest of the problems in the book. Or, if you prefer, save all three tests until you feel confident answering questions with no time pressure — then start the clock running and see how you do.

What Should You Study? Knowing What's on the ACT

The ACT covers six overall topics in math: pre-algebra, elementary algebra, intermediate algebra, coordinate geometry, plane geometry, and trigonometry. In this section, I break down all these topics into manageable bits and discuss the individual skills included in each. Parts 2, 3, and 4 (Chapters 4 through 13) cover this material in depth, with plenty of example questions and practice problems.

Taking care of the basics in pre-algebra

Pre-algebra includes a variety of topics that prepare you for algebra. In this section, I discuss the specific pre-algebra skills that show up most on the ACT. And in Chapter 4, I focus on these types of questions, providing plenty of example questions and showing you how to answer them.

Basic arithmetic

You obviously need to know the four operations: addition, subtraction, multiplication, and division. You also want to feel comfortable working with negative numbers, fractions, and decimals. I cover some of this material in this book, but if you feel that you need a more thorough review, pick up *Basic Math and Pre-Algebra For Dummies* (Wiley) by yours truly.

Number sequences

A *number sequence* is a list of numbers arranged in a pattern. Here's an example:

2, 5, 8, 11, 14, 17...

In this case, each number in the sequence is 3 greater than the number before it. An ACT question may ask you to find the next number or a missing number in a number sequence.

Factors and multiples

When one natural number is divisible by another, the smaller number is a *factor* of the greater number, and the greater number is a *multiple* of the smaller number. For example, 12 is divisible by 4, so

- 4 is a factor of 12.
- 12 is a multiple of 4.

To answer an ACT question, you may need to find all the factors of a number or the greatest common factor or the least common multiple among several numbers.

Fractions and Decimals

You probably first learned about fractions and decimals for the first time all the way back in fourth or fifth grade. If you're like many students, fractions and decimals can still slow you down or, in some cases, derail you entirely from getting the right answer.

At a minimum, you should know how to add, subtract, multiply, and divide fractions and decimals - either by hand or using your calculator. To add and subtract fractions with different denominators, you may also need to know how to find a common denominator. And to complete a problem, you may need to simplify fractions or change an improper fraction to a mixed number.

Percents, ratios, and proportions

Like fractions and decimals, *percents* are a mathematical way of representing part of a whole. For example, 50 percent of something is half of it. A *ratio* is a mathematical comparison. For instance, if you have twice as many brothers as sisters, the ratio of brothers to sisters is 2 to 1, or 2:1. A *proportion* is an equation using two ratios. ACT questions may ask you to calculate something using percents and ratios, or you may have to set up a proportion to answer a question.

Powers (exponents) and square roots (radicals)

When you take a number to a *power*, you multiply that number by itself repeatedly. For example: 3^4 (read *three to the fourth power*) $= 3 \times 3 \times 3 \times 3 = 81$. In this case, 3 is the *base* (the number multiplied) and 4 is the *exponent* (the number of times the base is multiplied).

And when you take a *root* (also called a *radical*) of a number, you find a result that can be multiplied by itself repeatedly to produce the number you started with. The most common root is the *square root* — a result which, when multiplied by itself, produces the number you started with. For example, $\sqrt{25}$ (read *the square root of 25*) $= 5$ because $5 \times 5 = 25$.

Powers and square roots are common math operations, and they show up a lot on all sorts of ACT questions.

Moving on to elementary algebra

Elementary algebra is essentially the algebra that's covered in an Algebra I class. In this section, I go over the highlights of what skills the ACT expects you to remember and work with when answering questions. Chapter 5 covers these topics in greater detail with lots of examples.

Evaluating, simplifying, and factoring expressions

An *expression* is any string of numbers and symbols that makes mathematical sense. In algebra, you can do three common things with expressions:

- **Evaluate:** To *evaluate* an expression, you plug in the value of each variable and change the expression to a number. Be sure to follow the order of operations (exponents in the order they occur left to right, multiplication and division in the order they occur left to right, and addition and subtraction in the order they occur left to right). For example, here's how you evaluate the expression $5x + 7$, given that $x = 4$:

  $$\begin{aligned} &5x + 7 \\ &= 5(4) + 7 \\ &= 20 + 7 \\ &= 27 \end{aligned}$$

- **Simplify:** To *simplify* an expression, you remove parentheses and combine like terms to make the expression more compact. For example, here's how you simplify the expression $3(x + 6) + 2x$:

  $$\begin{aligned} &3(x + 6) + 2x \\ &= 3x + 18 + 2x \\ &= 5x + 18 \end{aligned}$$

» **Factor:** To *factor* an expression, you find a factor that's common to each term in the expression and pull it out of the expression using parentheses. For example, here's how you factor $2x$ out of the expression $6x^2 - 10x$:

$$6x^2 - 10x$$
$$= 2x(3x - 5)$$

Easier ACT questions may ask you to simply evaluate, simplify, or factor an expression. More difficult questions may require you to use these skills to handle more complex calculations.

Solving equations with one or more variables

Solving equations is the main point of algebra. You solve an equation by isolating the variable (commonly x) while keeping the equation in balance — that is, by making sure that in each step, you apply the exact same operation to both sides of the equation. Here are a few types of equations you need to know how to solve on the ACT:

» Equations with fractions (rational equations), such as $\frac{2x}{5} = \frac{x+1}{4}$

» Equations with square roots (radicals), such as $\sqrt{5x+3} - 2 = x$

» Equations with absolute values, such as $|3x - 6| = 10$

» Equations with variables in the exponent, such as $8^{x-1} = 16$

Typically, an equation with more than one variable, such as $ab + c = 10$, can't be solved for a number. However, you can solve an equation with more than one variable in terms of the other variables in the equation. For example, here's how you solve this equation for b in terms of a and c:

$$ab + 3c = 10$$
$$ab = 10 - 3c$$
$$b = \frac{10 - 3c}{a}$$

An ACT question may ask you to solve an equation in terms of other variables. Additionally, this skill is useful when working with math formulas.

Data and graphs

A *graph* is a visual representation of data. Common graphs include bar graphs, pie charts, line graphs, and pictograms. Graph reading is a basic but essential skill that you need for the ACT. A typical question may ask you to identify specific data given in a graph, or you may need to pull this data as a first step in a more complex calculation.

Basic statistics and probability

Statistics is the mathematical study of real-world information called *data sets* — lists of numbers that are objectively observed and recorded. Three common operations used on data sets are three types of averages called the *mean*, the *median*, and the *mode*. On the ACT, you need to know how to calculate all three.

Statistics and probability

Probability measures the mathematical likelihood that an event will occur. On the ACT, you may need to calculate simple, compound, or conditional probability.

Focusing on intermediate algebra

Intermediate algebra is the focus of a high school Algebra II class. In this section, I outline the essential intermediate algebra skills you need to be successful on the ACT. Later on, in Chapter 8, you can gain a solid understanding of this material.

Taking a look at inequalities

An *inequality* is a statement telling you that two math expressions aren't equal. On the ACT, inequalities come in four basic varieties:

- Greater than (>)
- Less than (<)
- Greater than or equal to (≥)
- Less than or equal to (≤)

You solve inequalities using the same algebra rules you would use to solve equations — with the exception of a couple of twists (flip to Chapter 7 for details). The solution to an inequality is typically a range of answers expressed as a simpler inequality.

Working with systems of equations

A *system of equations* is made of two equations that are simultaneously true. On the ACT, a system of equations usually is limited to two variables. For example, take a look at this system:

$$3x + y = 10$$
$$x - 5y = -4$$

You can solve a simple system of equations by the *substitution method*, isolating a variable in one equation and then plugging its equivalent into the other equation. For a more complicated system of equations, use the *elimination* (or *combination*) *method* by either adding or subtracting the two equations and solving the equation that remains.

Understanding direct and inverse proportionality

When two values, x and y, are *directly proportional*, a value, k, makes the following equation true:

$$\frac{x}{y} = k$$

Values that are directly proportional tend to rise and fall together. For example, when one value doubles, the other value also doubles.

When two values, x and y, are *inversely proportional*, a value, k, makes the following equation true:

$$xy = k$$

Values that are inversely proportional tend to rise or fall opposite of each other. For example, when one value is multiplied by 3, the other value is divided by 3.

Examining quadratic equations

A *quadratic equation* is an equation in the form $ax^2 + bx + c = 0$. You can solve a quadratic equation either by factoring or by using the quadratic formula:

$$x = \frac{-b \pm \sqrt{b^2 - 4ac}}{2a}$$

The ACT almost certainly will have several questions that require you to work with quadratic equations.

Finding information about functions

A *function* is a mathematical connection between two values. Usually, the values are an input variable, x, and an output variable, y. In a function, when you know the value of x, the value of y is determined.

Typical ACT questions may ask you to use functions as models, to work with functional notation $f(x)$, to simplify the composition of two functions, to find the inverse of a function, or to find the domain or range of a function.

Working with coordinate geometry

Coordinate geometry is geometry that occurs on the xy-graph. This topic overlaps with material introduced in both Algebra I and Algebra II classes. Here, I give you an overview of the basic information from coordinate geometry that you need to review to do well on the ACT. I go over these ideas in greater detail in Chapter 9.

Graphing linear functions

A *linear function* is any function of the form $y = mx + b$. For example:

$$y = 3x + 5 \qquad y = \frac{5}{6}x - \frac{1}{3} \qquad y = -x$$

Linear functions, which produce a straight line when graphed, are common on the ACT. Some of the skills you need to feel comfortable with include mastering the distance and midpoint formulas, finding the slope of a line, using the slope-intercept form to solve problems, and working with parallel and perpendicular lines.

Recognizing quadratic functions

A *quadratic function* is in the form $y = ax^2 + bx + c$. For example:

$$y = 2x^2 + 11x + 9 \qquad y = x^2 + 4x + 4 \qquad y = x^2 - 1$$

On the graph, a quadratic function produces a *parabola* — a curve that looks roughly like an arch (or a U). On the ACT, a question may ask you to pair up a quadratic function with its graph. More difficult questions may require you to find the axis of symmetry or the vertex of a parabola or to solve a quadratic inequality.

Transforming functions

A *transformation* of a function is a small change that affects that function in a predictable way. Typical transformations include reflections across the x-axis and y-axis as well as vertical and horizontal shifts. An ACT question may ask you to compare two similar functions and select the equation that transforms one into the other. Or a question may provide a function and a transformation and ask you to produce the resulting graph.

Grappling with higher-order polynomial functions and circles

More difficult ACT questions may include higher-order polynomials, such as cubic equations of the form $y = ax^3 + bx^2 + cx + d$, and graphs of circles. These questions are rather uncommon and require only a basic familiarity with the concepts.

Reviewing plane geometry

Plane geometry is the focus of a typical high school geometry class. In this section, I discuss the geometry that you're likely to see on the ACT. Chapter 11 gives you a complete review of these topics.

Lines and angles

One common type of ACT question presents you with a figure that contains lines and angles and then asks you to find the value of a given angle. To answer this type of question, you need to know how to measure right angles, vertical angles, supplementary angles, the angles in a triangle, and the angles that result when two lines are parallel.

For example, an ACT question may show you a figure with some angles labeled and ask you to find the measure of an unlabeled angle. Or it may ask you to identify a pair of angles that are equal in measure.

Triangles

Virtually every ACT includes several questions about triangles. You may need to find the area of a triangle given the height and the base, use the Pythagorean theorem to work with right triangles, or work with the most common types of right triangles, such as the 3-4-5 triangle.

An ACT question may ask you to find the area of a triangle given the measurements of its height and base, or, turning this question around, it may ask you to find the height given the length of the base and the area of the triangle. ACT questions involving right triangles may ask you to identify the length of one side of a right triangle, given information about the other sides.

Quadrilaterals

A *quadrilateral* is a four-sided polygon. Basic quadrilaterals that you may encounter on the ACT include squares, rectangles, parallelograms, and trapezoids. You need to know how to find the area of all these, and, more generally, you must feel comfortable working with the formulas for these areas.

For example, an ACT question may give you the perimeter of a rectangle with additional information and ask you to find the area. Or it may give you information about some aspects of a parallelogram — such as its height and area — and ask you to calculate the length of its base.

Circles

Circles are quite common on the ACT. You need to know the formulas for finding the diameter, area, and circumference of a circle given its radius. Additionally, you should be able to work with tangent lines, arc length, and chords of circles.

An ACT question may ask you to find the circumference of a circle given its area. More difficult ACT questions may require you to combine other geometry formulas to measure the area of a triangle with one side that's tangent to a circle or a chord of a circle.

Solid geometry

Solid geometry deals with geometry that occurs in three-dimensional space. A basic ACT question may require you to find the volume of a cube or box (rectangular solid). More advanced questions may ask you to work with more complicated solids, such as spheres, prisms, cylinders, pyramids, and cones.

Dealing with trigonometry and other advanced topics

The ACT includes questions about a few advanced math topics, including trigonometry. In this section, I go over these topics to make sure you're prepared for them. For further details, check out Chapter 12.

Trigonometry

Trigonometry is the mathematics of triangles — most commonly right triangles. ACT questions cover basic trig information. For instance, you need to know how to find the six trig ratios of a triangle in terms of the opposite side, adjacent side, and hypotenuse. More advanced trig concepts deal with radian measure, graphs of trig functions, and some basic trig identities.

Matrices

A *matrix* is a grid of numbers with both a horizontal and a vertical dimension. Virtually every ACT has a question that asks you to recall basic information about matrices, such as adding or subtracting matrices, multiplying a matrix by a constant, or working with the determinant of a 2-by-2 matrix.

Logarithms

A *logarithm* is the inverse form of an exponent. Not every ACT includes a question about logarithms, but if you encounter this type of a question, knowing how to convert a logarithmic equation into an exponential equation is particularly helpful.

Imaginary and complex numbers

An *imaginary number* includes the value i where $i^2 = -1$. A *complex number* is a number of the form $a + bi$. ACT questions about these types of numbers aren't usually difficult. In fact, some basic information can help you to answer them.

You're Off the Hook: Discovering What the ACT Doesn't Cover

The ACT math test covers most of the topics you're likely to find in a basic high school math curriculum. In fact, it's more advanced than the SAT in its range of math topics. Fortunately, even the ACT doesn't require you to know *everything* about math. Here are three easily identifiable areas of math that the ACT doesn't cover:

- **Ellipses and hyperbolas:** The equations and graphs for ellipses and hyperbolas, often part of an Algebra II or a pre-calculus class, aren't present on the ACT.
- **The value *e* and natural logarithms:** In a pre-calculus course, your teacher introduces you to the value *e* and its inverse function, the natural log. Both of these areas are essential for calculus, but you don't need to worry about them for the ACT.
- **Calculus and beyond:** More and more high school students are taking one or even two years of calculus and other advanced math. On the ACT, you definitely don't have to worry about limits, derivatives, integrals, or any other advanced concepts that you encounter in a calculus class.

This information comes as good news for most students. If you're currently taking an advanced math class, of course, you still need to study to maintain your grades. (You don't want to get a 36 on your ACT and then be rejected from your first-choice college because of low grades, right?) On the plus side, you may find many ACT questions easier than last night's homework.

On the other hand, if your goal in life is to avoid as much math as possible going forward, then as the saying goes "You may already be a winner!" That is, if you've passed high school Algebra I and II, and Geometry, you should be in reasonably good shape.

In either case, use this book to review the topics you're shaky on, solidify these skills with practice problems, and then take the practice tests in Part 5. You'll increase your confidence going into the ACT and come away from the test with a score you can be proud of.

IN THIS CHAPTER

» Taking the test in two passes

» Knowing when to guess an answer

» Understanding multiple-choice questions

» Using your calculator to your advantage

Chapter 2 Boosting Your Test-Taking Skills

Some folks say that the only thing tests measure is how well you take a test. Such news would be great, because then you could skip all the math stuff in this book and head directly to tips such as "always guess Choice (D)." But guess what? That tactic will only get you so far!

Even so, before you take any test, you do need to know a few strategies that relate specifically to that test. In this chapter, I focus on some basic facts about ACT math and show you how best to approach the test to take advantage of these facts.

To begin, I discuss the strategy of tackling the math section in two passes — the first to handle the easy questions and the second to focus on the more difficult ones. I also discuss a few strategies about guessing on the ACT and provide you with a few ways to approach multiple-choice math questions. Finally, I end the chapter with some advice on selecting a calculator, including some tips for using the one you select.

Two-Timing the Test: Taking Two Passes to Answer Questions

On the ACT math test, you have 60 minutes to answer 60 questions. So you have roughly 1 minute per question. Every question you answer correctly is worth 1 point toward your raw score.

But not all ACT math questions are created equal. Generally speaking, the questions increase in difficulty as you proceed from Question 1 to Question 60. Here's the general breakdown of difficulty:

» **Easy:** Questions 1 through 20

» **Medium:** Questions 21 through 40

» **Hard:** Questions 41 through 60

You should obviously start the test at Question 1 and proceed in order as much as you can. However, every student is different, so you may find some questions along the way that are more difficult for you than later questions. So if you read a question and don't have a clue how to answer it, feel free to jump over it. The next question may be easier for you to answer. I suggest you use the tried-and-true strategy of taking two passes over the ACT:

» **Pass #1:** Start with Question 1 and work your way forward, answering questions that look relatively quick and easy and jumping over those that look difficult or time-consuming. Don't guess at this stage.

» **Pass #2:** After you've answered all the quick and easy questions you can, circle back to the first question you skipped over and work your way forward to the end again.

This strategy maximizes the number of questions you can answer with confidence. It also helps you save time for the tough questions, which usually take more than 1 minute to solve.

TIP

In my humble opinion, every ACT math section includes a few questions that are practically *begging* for you to skip over them. For example, you may consider passing over questions that

» Are very long and wordy

» Seem purposely confusing and don't make a lot of sense even the second time you read them

» Have large or complicated numbers that will involve long or difficult calculations

Of course, not every problem with the preceding characteristics is as difficult as it looks. But as you run across problems like this, feel free to jump over them — even on Pass #2. If you have time at the end of the test, you can always try to pick off a few of these questions. But if you're going to skip questions, you may as well skip these hairy beasts.

REMEMBER

No penalty exists for guessing on the ACT, so be sure to answer all 60 questions before your time runs out. I discuss this point further in the next section, "To Guess or Not to Guess."

To Guess or Not to Guess

On the ACT, you don't lose points from your raw score when you fill in a wrong answer. So strategically you should fill in *every* answer, even if you have to make a wild guess.

Of course, you don't want to guess on questions that you may be able to answer correctly — especially among the earlier questions, which tend to be easier. And keep in mind that an educated guess is always better than a wild guess. So whenever possible, rule out answers that you know are wrong. Keep track of these wrong answers by crossing them out in your test booklet.

Don't guess at any answers while you're still on the first pass (see the previous section, "Two-Timing the Test: Taking Two Passes to Answer Questions," where I discuss tackling the test in two separate passes). Instead, begin guessing on your second pass. At this point, if you can confidently rule out a couple of answers but don't know how to proceed with a question, you can save time by guessing at the answer and moving on to the next question.

TIP

Whenever possible, keep track of the questions that you guess on. If you have time at the end of the test — or if you have an unexpected brainstorm — you can revisit these questions.

Finally, keep close track of your time. When your 60 minutes are almost up, take a moment to guess at all the remaining answers — don't leave any blank. With a bit of luck, you may pick up a few additional points on some of these questions.

WARNING

After your time is up on any part of the test, the ACT rules state that you may *not* return to that part, even to fill in guesses. If you get caught doing this, you could be expelled from the test with no score and no refund. 'Nuff said!

Answering Multiple-Choice Questions

The math section of the ACT comprises 60 multiple-choice questions. Each question provides five possible answers. You likely have been taking standardized tests for most of your life as a student, so you're probably already familiar with this type of question. However, math teachers generally don't use multiple-choice questions when assigning homework or testing. So in this section, I provide a few strategies for approaching multiple-choice math questions.

Considering the five answer choices

Every multiple-choice question gives you a little extra information, because you know the correct answer must be one of the five choices given. Always take a moment to notice these answer choices, because they may guide you as you work on solving the problem. The following example shows you how you can rely on answer choices to correctly solve a problem.

EXAMPLE

If $j^2 - 14j + 48 = 0$, which of the following shows *all* of the possible values of j?

(A) -6

(B) 8

(C) 6, 8

(D) −6, 8

(E) −6, −8

You can solve the equation $j^2 - 14j + 48 = 0$ by factoring. (I show you the details of factoring in Chapter 5 and explain how to apply this technique to solving quadratic equations in Chapter 8.) In this case, every value in each of the five answers includes either 6 or 8 (give or take a minus sign), so you have a head start on the factoring:

$$\begin{aligned} j^2 - 14j + 48 &= 0 \\ (j\ \ 6)(j\ \ 8) &= 0 \end{aligned}$$

At this point, you only need to fill in the signs (+ or −) inside the parentheses. Because 48 in the original equation is positive, the two signs must be the same (either both + or both −). And because −14 is negative, at least one of the signs is negative. Therefore, both signs are negative:

$$(j-6)(j-8) = 0$$

Now you can solve this equation by breaking it into two separate equations:

$$\begin{aligned} j - 6 &= 0 \qquad & j - 8 &= 0 \\ j &= 6 & j &= 8 \end{aligned}$$

Thus, the correct answer is Choice (C).

Plugging and playing

Multiple-choice questions give you an opportunity to arrive at the correct answer by plugging in the answer choices and solving. Note that plugging in answers can be a little time-consuming, so if you can find a better way to solve the problem, go for it. But when you get stuck, this tactic gives you a chance at answering questions that you really aren't sure how to solve. Consider the following example.

EXAMPLE

If $\sqrt{5x+1}-1=x-2$, then $x=$

(A) 4

(B) 5

(C) 6

(D) 7

(E) 8

You may or may not know how to solve this type of equation (flip to Chapter 5 to see how it's done). And in any case, solving it may be time-consuming. So you can try to plug in each possible answer for x to see which one works. Start with Choice (A) and plug in 4 for x:

$$\begin{aligned}\sqrt{5(4)+1}-1 &= 4-2\\ \sqrt{21}-1 &= 2\\ \sqrt{21} &= 3 \quad \textit{Wrong!}\end{aligned}$$

This answer choice is obviously wrong, because 21 isn't a square number. Therefore, $\sqrt{21}$ is irrational and doesn't equal 3. In fact, this wrong answer choice may suggest a way to save even more time: The reason this answer is wrong is that the value of $\sqrt{5x+1}$ evaluates to an irrational number, which messes up the equation. So $\sqrt{5x+1}$ has to be a rational number, which means $5x+1$ must be a square number. Try testing Choices (B) through (E) in this way, keeping in mind that you're looking for a value of x that makes $5x+1$ a square number:

(B) $: 5(5)+1=25+1=26$ not a square number

(C) $: 5(6)+1=30+1=31$ not a square number

(D) $: 5(7)+1=35+1=36$ a square number!

(E) $: 5(8)+1=40+1=41$ not a square number

Only one value produces a square number, so the correct answer is Choice (D). You can verify this by plugging in 7 for x:

$$\begin{aligned}\sqrt{5(7)+1}-1 &= 7-2\\ \sqrt{36}-1 &= 5\\ 6-1 &= 5\end{aligned}$$

Some questions ask you for the *greatest* or *least* number that has a certain property. These questions provide a great opportunity to test answers individually until you find the correct one. Consider the following strategies:

- When looking for the *lowest* or *least* value, begin with the lowest number and work your way up.
- When looking for the *greatest* or *highest* value, begin with the greatest number and work your way down.

The following example illustrates this strategy.

EXAMPLE

What is the least common denominator when adding three fractions with denominators of 6, 9, and 16?

(F) 60

(G) 120

(H) 144

(J) 240

(K) 288

Because you're looking for the *least* common denominator, you can find the correct answer by testing numbers and ruling out wrong answers, starting with the lowest number.

Begin by testing to see whether 60 is divisible by 6, 9, and 16:

$$60 \div 6 = 10$$
$$60 \div 9 = 6 \text{ r } 6 \quad \text{not divisible}$$

So Choice (F) is wrong. Now test 120:

$$120 \div 6 = 20$$
$$120 \div 9 = 13 \text{ r } 3 \quad \text{not divisible}$$

So Choice (G) also is wrong. Next, test 144:

$$144 \div 6 = 24$$
$$144 \div 9 = 16$$
$$144 \div 16 = 9$$

So Choice (H) is the correct answer. By the way, before moving on, notice that 288 is also divisible by all three denominators. However, Choice (K) is wrong because the question asks for the *least* common denominator, which is why you started plugging in the lowest numbers first.

Calculating Your Way to Success: Calculators and the ACT

If you hate doing long division as much as I do, you're probably glad to hear that calculators are allowed on the ACT. In this section, I answer a few basic questions about calculators.

When should you use a calculator?

A calculator is a great tool for solving problems more quickly than you can either in your head or using a pencil and scratch paper. At the same time, however, you want to avoid overusing it for calculations that you can easily and accurately do in your head.

When you take a practice test, notice how you use your calculator. Do you almost forget about it? If so, here are a few tips to train yourself to use the calculator to your advantage:

- Look over the practice test and see whether you performed any long calculations that a calculator could have saved you time with.

- Be sure you know how to use your calculator. If you aren't familiar with it, you're less likely to use it when appropriate. Check out the tips for calculator use later in this chapter.
- Consider upgrading to a calculator that can handle things like square roots, powers, fractions, or other types of math that you need help with on the ACT. I provide a few ideas later in this section.

On the other hand, do you use your calculator for just about every problem? If so, consider these tips for backing off a bit:

- Notice a few places (especially at the beginning of the test) where you may have done an easy calculation more quickly without your calculator.
- If you overuse your calculator for simple arithmetic, such as for multiplying 6×8 or dividing $14 \div 2$, you probably can benefit from beefing up your basic math skills. Check out Chapter 3 for a list of must-know math skills for the ACT.
- If you second-guess yourself too much — that is, you do basic math *correctly* without a calculator but then doubt the answer and check it to be sure — you probably need to convince yourself that you're on track. Start to notice how often you check simple math and find that your answer was already correct. If you're almost always correct, you're probably spending time using your calculator that could be used in a better way.

What kind of calculators can you use?

The calculators that are allowed on the ACT are divided into these categories:

- **Basic calculators:** You can buy this type of calculator for less than $10 in almost any store that sells stationery. Basic calculators are perfect for tedious calculations, such as adding or multiplying big numbers, but they're simply not adequate for the ACT. If you're currently using this kind of a calculator, I strongly suggest that you consider upgrading to a scientific or graphing calculator.
- **Scientific calculators:** This type of calculator typically costs more than $10, but you get a lot of functionality not found on a basic calculator. Depending on the model, a scientific calculator usually includes exponents, square roots, logarithms, trig functions, a reciprocal function, and lots of other stuff that may come in handy on the ACT.
- **Graphing calculators:** A graphing calculator has all the bells and whistles of a scientific calculator, *plus* a larger screen for visual display of graphs and tables. If you're thinking of upgrading to a graphing calculator from either of the other two types, consider this: The main advantage you gain is directly related to your proficiency with these visual elements. So plan to spend at least four or five hours practicing with your new toy, creating input-output tables for functions, graphing lines and parabolas, and exploring other related visual options. If you're not convinced you're really going to practice, you may as well save your money. Stick with a scientific calculator, which should serve you well enough.

What kind of calculators can't you use?

WARNING

The elders of the ACT weren't born yesterday, which is how they got to be elders! So don't try to pull a fast one on them. Even though you're allowed to use a calculator on the test, you may *not* use a calculator that includes any of the following features:

- **Texting and Internet access:** Sure, your iPhone (or iPad or laptop) may have a calculator function, but this function doesn't *make* it a calculator. It also has lots of other fancy capabilities that aren't allowed on the ACT. Obviously the elders don't want you texting your genius Uncle Roy at MIT or finding answers using Photomath (a great app otherwise, by the way) if you get stuck on a question.
- **Talking or other weird noises:** If your calculator makes noise and disturbs people, the monitors may separate you from it for the duration of the test. Of course, that separation wouldn't be good for your test score.
- **Electrical access:** I can't guarantee that your testing site will have a place to plug in a calculator that requires power. Even though you may get lucky, your best bet is to bring a battery-powered calculator (along with a fresh set of batteries).

The point here is that the ACT elders are traditionalists. So if you stray much beyond the old-fashioned scientific and graphing calculators, you may run into problems.

How do you use your calculator?

After you purchase your calculator, don't let it sit in its impenetrable plastic packaging until the night before the ACT. Use it for at least one practice test so you can get the feel of where the important keys are. At a minimum, make sure you know how to enter the following:

- **Negative numbers:** Scientific calculators may have a special key to enter a negative number. This key may be distinct from the minus sign used for subtraction. Find this key and test it by calculating a few things, such as $-1-5=-6$
- **Parentheses:** When you enter complicated calculations, you may need parentheses. The most common example of this is when you enter a fraction such as $\frac{2+7}{4-1}$. Enter this fraction as $(2+7)/(4-1)$ and make sure you get the answer 3.
- **Pi (π):** On most scientific calculators, the π function is simple to use, requiring only one or two key strokes. Locate it and calculate 10π — the answer should be about 31.4.
- **Square roots:** On many popular scientific calculators, the square root function doesn't have its own key. Instead, it's often the 2nd function on the x^2 key used for squaring a number. This manipulation isn't complicated, but you want to know it cold before you take your ACT. Make sure you can calculate some square roots, such as $\sqrt{196}=14$.
- **Trig functions:** Truthfully, you may never need to use the sin, cos, or tan keys. Even so, be prepared. Locate them and calculate $\cos 0=1$.

  Also very important: Scientific calculators accept trig inputs in either degrees or radians, depending on which mode you choose. (Here's a quick test: Enter $\sin \pi$ and see what you get. If the answer is 0, you're in *radian mode*; if it's something weird, you're in *degree mode*.) Both modes work equally well, so decide whether you like working with degrees or radians best and make sure your calculator is set for this mode.

If you have a fancy graphing calculator, here are some of the useful features that are worth checking out:

- **Input-output tables:** This feature allows you to enter a function such as $y=2x-3$, and then the calculator builds a table showing the resulting *y*-values (given $x=1$, $x=2$, and so on).
- **Solving equations:** You won't want to miss this time-saving feature. Be sure to enter equations carefully, using parentheses as needed, especially for complicated fractions.

» **Graphs:** This feature allows you to enter a function and view the resulting graph. More advanced features allow you to solve equations — including quadratic equations — using a graph.

TIP

For more tips on using a graphing calculator, check out *TI-83 Plus Graphing Calculator For Dummies* (Wiley) and *TI-89 Graphing Calculator For Dummies* (Wiley), which are both written by C. C. Edwards. Finally — and I know that in the fun department, this ranks somewhere between comparison shopping for snow tires and cleaning the lint filter on your dryer — you can learn a lot about your calculator from that little manual that comes along with it. Don't just toss it aside!

IN THIS CHAPTER

» Putting your basic math skills to the test on the ACT

» Considering which formulas to remember and understanding how to use them

» Jotting and sketching your way to success with word problems

Chapter 3
Discovering Some Problem-Solving Strategies

Every math test — especially the math portion of the ACT — requires some basic problem-solving skills. So in this chapter, I outline these skills. First, I discuss the basic math skills you should remember from before high school: the four basic operations (addition, subtraction, multiplication, and division), negative numbers, the order of operations (which is affectionately known by the mnemonic PEMDAS), and fractions. Next, I list the math formulas, along with some tips on how to use them, that appear most on the ACT. Finally, I discuss a few basic techniques for working with every student's favorite type of math question: word problems.

REMEMBER

You can use the skills I review in this chapter throughout the entire math portion of the ACT. No matter what the topic is, basic math problem solving can always help you out.

Identifying Basic Math Skills You Need to Know

The ACT tests your understanding of the math that most students take in high school: pre-algebra, algebra, geometry, and a bit of trigonometry. In this book, I focus specifically on these topics. However, to do well on the test, you also need a good grounding in the math basics that are taught in the earlier grades.

So in this section, I give you a look at these basic skills that you need to remember. Make sure you feel confident with this material. If you're concerned that some of these skills may hang you up on the test, I recommend that you pick up *Basic Math and Pre-Algebra For Dummies* (Wiley) for thorough and detailed explanations of these skills. And for additional practice, you also can grab *Basic Math and Pre-Algebra Workbook For Dummies* (Wiley). Both of these are written by me, so I'm hoping you'll like them and maybe even buy them.

Using the four basic operations

The four basic operations — addition, subtraction, multiplication, and division — are the underpinning of everything in math. The ACT doesn't directly test these skills, but you need to use them on virtually every question you face.

Fortunately, you can use your calculator, so you won't need to do complicated calculations — such as long division — with paper and pencil. On the other hand, you don't want to waste time doing every tiny calculation on your calculator.

REMEMBER

At a minimum, make sure you know the following:

- The addition table up to $9+9=18$
- All corresponding subtraction inverses up to $18-9=9$
- The multiplication table up to $9\times 9=81$
- All corresponding division inverses up to $81\div 9=9$

If you're a little sketchy (and, be honest, who isn't a little sketchy about 6×9 and 7×8?), a couple of days working with flash cards could be time well spent.

Staying positive with negatives

Negative numbers aren't all that difficult, but they can be confusing — especially when you're trying to move quickly on a test. Here are a few basic points to remember:

- **When you negate a number, you change its sign (either from + to –, or from – to +).** For example:

  The negation of 5 is -5. The negation of -8. is 8.

- **When you subtract a larger number from a smaller one, the result is negative.** For example:

  $3-7=-4$ *because* $7-3=4$

- **Adding a negative number is the same as subtracting a positive number.** For example:

  $6+(-1)=6-1=5$

- **When you subtract a negative number, the minus signs cancel out to become addition.** For example:

  $9-(-2)=9+2=11$

- **When you multiply or divide a positive number by a negative number (in either direction), the result is negative.** For example:

  $3\times(-6)=-18$ $\quad 8\div(-2)=-4$

  $-5\times 10=-50$ $\quad -20\div 4=-5$

- **When you multiply or divide a negative number by a negative number, the minus signs cancel out and the result is positive.** For example:

  $-9\times(-3)=27$ $\quad (-45)\div(-5)=9$

Making peace with PEMDAS

When you use more than one operation at a time, ambiguities arise that may cause you to arrive at the wrong answer. For example, suppose you have the following problem:

$$3+6\times 2=?$$

You might potentially approach this problem in two different ways. You could either do the addition first and the multiplication second or vice versa. But each of these approaches would give you a different answer:

$$(3+6)\times 2=9\times 2=18 \quad 3+(6\times 2)=3+12=15$$

REMEMBER

To avoid this problem, mathematicians have standardized the *order of operations* (also called *order of precedence*), which is a set of rules telling you the order in which to break down large problems. Here is the order of operations that you need for all math on the ACT (you can remember the order with the mnemonic PEMDAS):

1. Parentheses
2. Exponents
3. Multiplication and Division
4. Addition and Subtraction

REMEMBER

For each of these four steps, work from left to right. Check out the following example problem to see how these steps work:

$$(3+7)^2-(2\times 4)\times 2^3$$

To begin, evaluate what's inside the parentheses, from left to right:

$$=10^2-8\times 2^3$$

Next, evaluate the exponents, from left to right:

$$=100-8\times 8$$

Now evaluate the multiplication:

$$=100-64$$

Finally, evaluate the subtraction:

$$=36$$

You probably won't need to evaluate anything quite this complicated on the ACT, but here's the point: If you remember the order of operations, you won't get into trouble.

WARNING

A common error when applying PEMDAS is to mistakenly believe that multiplication should be done before division, or that addition should be done before subtraction. In fact, multiplication and division are *equivalent* in order, as are addition and subtraction. This table shows the four levels of PEMDAS ordering:

Order	Letter	Operation
1st	P	Parentheses
2nd	E	Exponents
3rd	MD	Multiplication and division
4th	AS	Addition and subtraction

A couple of quick examples should clarify this. For example, in the expression $20 \div 2 \times 5$, the division sign comes before the multiplication sign, so *divide first* and *multiply second*:

$$20 \div 2 \times 5 = 10 \times 5 = 50$$

Similarly, in the expression $14 - 4 + 6$, the minus sign comes before the plus sign, so *subtract first* and *add second*:

$$14 - 4 + 6 = 10 + 6 = 16$$

Here's a final example that includes all four basic operations:

$$30 - 18 \div 2 \times 3 + 5$$

Begin by doing the two MD operations (multiplication and division) from left to right, so divide first and multiply second:

$$= 30 - 9 \times 3 + 5 = 30 - 27 + 5$$

To complete the evaluation, do the two AS operations (addition and subtraction) from left to right, so subtract and then add:

$$= 3 + 5 = 8$$

One last word: You may have seen videos on YouTube or elsewhere dramatically claiming that "PEMDAS is wrong" or "PEMDAS is broken." Most of these viral controversies are easily remedied when, in algebra, students move away from the symbols × and ÷ in favor of less ambiguous notation. In any case, relax: While the ACT may require you to solve problems by applying the PEMDAS rules, it won't ask you to interpret intentionally confusing Internet memes.

Making friends with fractions

Many calculators allow you to work with fractions. If your calculator has this functionality, spend some time getting comfortable with the following basic fraction manipulations:

- Adding, subtracting, multiplying, and dividing simple fractions
- Simplifying fractions
- Increasing a pair of fractions to a common denominator
- Performing fraction-decimal conversions for common fractions

REMEMBER

If you don't know how to perform these functions, find a video online to explain how to do them. On the other hand, if your calculator doesn't have this capability (or you're not prepared to figure out how it works), make sure you can do these basic manipulations quickly on paper.

Getting Comfortable with Formulas

A mathematical *formula* is an equation that allows you to find a value when you know one or more related values. For example, if you know the base and height of a triangle, you can find the area of that triangle using this formula:

$$A = \frac{1}{2}bh$$

You can use this formula to find any value (the area, base, or height) as long as you have the other two values. For example, you can find the base of a triangle by plugging the values of the area and height into the equation and solving.

In this section, I include a list of all the formulas I cover in this book. This list provides one-stop shopping to make sure you know these formulas before you take the ACT. But simply memorizing formulas isn't enough, so I also discuss how to use the formulas in your arsenal.

REMEMBER

This section is a mere preview of formulas. Throughout the book, I discuss each formula in much greater detail, show you how to use it, and then provide practice problems that reinforce this information.

Reviewing the formulas you need to know

In this section, I compile all the formulas that appear throughout the rest of the book and tell you what chapter discusses each in depth. This is pretty much a need-to-know list for the ACT, so be sure you memorize the ones you're not familiar with. But having said that, I can assure you that this book provides plenty of practice understanding and using these formulas. And when you work with them for a while, you may find the memorizing part pretty much takes care of itself.

Elementary algebra

In Chapter 5, where I talk about elementary algebra, I don't introduce any formulas. However, elementary algebra often is necessary for finding a value in a formula, particularly when the value you're trying to find isn't isolated on one side of the formula. I discuss this idea in greater detail in the later section "Working with your arsenal of formulas."

Statistics and probability

I review statistics and probability in Chapter 6. In that chapter, I introduce the following two important formulas:

- **Arithmetic mean:** Calculate the *arithmetic mean* (or *average*) of a set of values as follows:

  $$\text{Arithmetic mean} = \frac{\text{Sum of values}}{\text{Number of values}}$$

- **Probability:** Calculate the *probability* that an event will happen as a value from 0 to 1 using the following formula:

  $$\text{Probability} = \frac{\text{Target outcomes}}{\text{Total outcomes}}$$

Intermediate algebra

Here are three useful formulas that make their way into the intermediate algebra discussion in Chapter 8:

- **Direct proportionality:** Two variables, x and y, are *directly proportional* when the following equation is true for some constant k:

  $$\frac{x}{y} = k$$

» **Inverse proportionality:** Two variables, x and y, are *inversely proportional* when the following equation is true for some constant k:

$$xy = k$$

» **The quadratic formula:** For any *quadratic equation* — that is, an equation of the form $ax^2 + bx + c = 0$ — the value of x is as follows:

$$x = \frac{-b \pm \sqrt{b^2 - 4ac}}{2a}$$

Coordinate geometry

In Chapter 9, the topic is coordinate geometry. Lots of useful formulas emerge there. Here are the ones I focus on:

» **Midpoint formula:** The formula for coordinates of the midpoint of a line segment between any two points (x_1, y_1) and (x_2, y_2) is:

$$\text{Midpoint} = \left(\frac{x_1 + x_2}{2}, \frac{y_1 + y_2}{2}\right)$$

» **Distance formula:** The formula for the length of a line segment between any two points (x_1, y_1) and (x_2, y_2) is:

$$\text{Distance} = \sqrt{(x_2 - x_1)^2 + (y_2 - y_1)^2}$$

» **Rise-run slope formula:** The formula for the slope of a line in terms of the rise and the run is:

$$\text{Slope} = \frac{\text{Rise}}{\text{Run}}$$

» **Two-point slope formula:** The formula for the slope of a line that includes the points (x_1, y_1) and (x_2, y_2) is:

$$\text{Slope} = \frac{y_2 - y_1}{x_2 - x_1}$$

» **Axis of symmetry of a parabola:** The *axis of symmetry* is the vertical line that divides a parabola down the middle. For a quadratic function $y = ax^2 + bx + c$, the formula for the equation of the axis of symmetry is:

$$x = -\frac{b}{2a}$$

» **Vertex of a parabola:** The *vertex* of a parabola is the point where the parabola changes directions. It's always either the lowest or the highest point on the parabola. Here's the formula for the coordinates of the vertex for the quadratic function $y = ax^2 + bx + c$:

$$\text{Vertex} = \left(-\frac{b}{2a}, -\frac{b^2 - 4ac}{4a}\right)$$

» **Formula for graphing a circle:** The formula for the equation of a circle of radius r centered at the point (h, k) is:

$$(x - h)^2 + (y - k)^2 = r^2$$

Plane geometry

Plane geometry, which I discuss in Chapter 11, is chock-full of useful formulas. If you can't commit all these formulas to memory and need to prioritize them, memorize everything up to the circumference of a circle. These are the must-know formulas for geometry. Then, if you have time, work your way forward from there.

- **Interior angles of a polygon** (the total of all angles inside the polygon):

  $$\text{Total interior angles} = 180\,(\text{Number of angles} - 2)$$

- **Area of a triangle:**

  $$A = \frac{1}{2}bh$$

- **Pythagorean theorem:**

  $$a^2 + b^2 = c^2$$

- **Area and perimeter of a square:**

  $$A = s^2$$
  $$P = 4s$$

- **Area and perimeter of a rectangle:**

  $$A = lw$$
  $$P = 2l + 2w$$

- **Area of a parallelogram:**

  $$A = bh$$

- **Area of a trapezoid:**

  $$A = \frac{b_1 + b_2}{2}h$$

- **Diameter, area, and circumference of a circle:**

  $$D = 2r$$
  $$A = \pi r^2$$
  $$C = 2\pi r$$

- **Arc length of a circle:**

  $$\text{Arc length} = \text{degrees}\ \frac{\pi r}{180} \quad \text{and} \quad \text{Arc length} = \text{Radius} \times \text{Radians}$$

- **Volume and surface area of a cube:**

  $$V = s^3$$
  $$A = 6s^2$$

- **Volume and surface area of a box:**

  $$V = lwh$$
  $$A = 2lw + 2lh + 2wh$$

- **Volume and surface area of a sphere:**

  $$V = \frac{4}{3}\pi r^3$$
  $$A = 4\pi r^2$$

- **Volume of a prism:**

  $V = A_b h$

- **Volume of a pyramid:**

  $V = \frac{1}{3} A_b h$

- **Volume of a cylinder:**

  $V = \pi r^2 h$

- **Volume of a cone:**

  $V = \frac{1}{3} \pi r^2 h$

Trigonometry and other topics

In Chapter 12, I discuss trigonometry and a few other advanced math concepts that may show up on the ACT. Here's a summary of the formulas I introduce in that chapter:

- **Six trig ratios:**
  - $\sin x = \frac{O}{H}$
  - $\cos x = \frac{A}{H}$
  - $\tan x = \frac{O}{A}$
  - $\cot x = \frac{A}{O}$
  - $\sec x = \frac{H}{A}$
  - $\csc x = \frac{H}{O}$
- **Formula for radians:**
  - $\frac{180}{\pi} = \frac{\text{degrees}}{\text{radians}}$
- **Five trig identities:**
  - $\sin x = \frac{1}{\csc x}$
  - $\cos x = \frac{1}{\sec x}$
  - $\tan x = \frac{1}{\cot x}$
  - $\tan x = \frac{\sin x}{\cos x}$
  - $\cot x = \frac{\cos x}{\sin x}$
- **Determinant:** The formula for the determinant of a 2-by-2 matrix $M = \begin{bmatrix} a & b \\ c & d \end{bmatrix}$:

  $\text{Det}(M) = ad - bc$

- **Expressing a logarithmic equation as an exponential equation:** The logarithmic equation $\log_b n = a$ is equivalent to the following equation:

  $b^a = n$

- **Formula for the imaginary number *i*:**

  $i = \sqrt{-1}$ and $i^2 = -1$

Working with your arsenal of formulas

Just as important as knowing the necessary formulas is knowing how to use them. The good news here is that in a certain sense, when you've seen one formula, you've seen them all. Every formula is simply an equation containing two or more variables. When you fill in all but one value, you can use algebra to solve for the missing value. In this section, I go over the most common skills you need for the ACT when working with formulas.

Solving for a missing value

Generally speaking, when you know all the values in a formula except for one, you can find the missing value. You do this by plugging in the values you know and then using algebra to find the remaining value. The following example illustrates this concept.

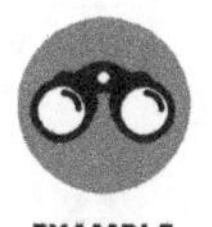

EXAMPLE

If the width of a rectangle is 4 and its perimeter is 28, what is its length?

(A) 7

(B) 8

(C) 9

(D) 10

(E) 11

Use the formula for the perimeter of a rectangle:

$$P = 2l + 2w$$

Plug in 4 for w and 28 for P:

$$28 = 2l + 2(4)$$

Solve for l:

$$28 = 2l + 8$$
$$20 = 2l$$
$$10 = l$$

Therefore, the correct answer is Choice (D).

Using two different formulas to solve one problem

To answer some questions, you may need to use two different formulas. Always start with the formula for which you can fill in all but one value, and then use this value to plug into the next formula. The following example illustrates this idea.

EXAMPLE

If the circumference of a circle is 9π, what is its area?

(A) 6π

(B) 18π

(C) 20.25π

(D) 40.5π

(E) 81π

The question gives you the circumference of the circle and asks for the area. So use the formula for the circumference first:

$$C = 2\pi r$$

Fill in 9π for C, and then solve for r:

$$9\pi = 2\pi r$$
$$4.5 = r$$

Now you can use the formula for the area of a circle:

$$A = \pi r^2$$

Plug in 4.5 for r, and then solve for A:

$$A = \pi r^2 = \pi(4.5)^2 = 20.25\pi$$

So the correct answer is Choice (C).

Working with unfamiliar formulas

When you know how to use a formula, you don't even need to know what the formula means. For example, here's a question that uses a formula for a trig identity that I discuss in Chapter 12:

EXAMPLE

If $\cos x = 0.8$ and $\tan x = -0.75$, then what is the value of sin x, given the formula $\tan x = \frac{\sin x}{\cos x}$?

(A) 0.6

(B) –0.6

(C) 0.75

(D) 0.8

(E) –0.8

If you haven't studied trigonometry, you may not have a clue what this question is asking. Even so, you probably know you can plug in 0.8 for cos x and –0.75 for tan x:

$$\tan x = \frac{\sin x}{\cos x}$$
$$-0.75 = \frac{\sin x}{0.8}$$

Now use algebra to solve for sin x. Simply multiply both sides of the equation by 0.8:

$$-0.6 = \sin x$$

Thus, the correct answer is Choice (B).

Sorting Through Word Problems

A *word problem* (also called a *story problem* or a *problem in a setting*) gives you information in words rather than in just equations and numbers. To answer an ACT word problem, you have to translate the provided information into one or more equations and then solve.

Throughout this book, I provide practice answering word problems that focus on specific math skills, such as algebra or geometry. In this section, I give you a few general tools for solving word problems.

Jotting down the numbers

You can solve some word problems fairly easily. Jotting down the numbers in the problem can be useful to help get you focused and moving in the right direction. The example word problems in this section show you how.

EXAMPLE

Seminar X brought in $700 in revenue and had 20 participants, each of whom paid the same amount. Seminar Y brought in $750 and had 15 participants, each of whom paid the same amount. How much more did each person pay for Seminar Y than Seminar X?

(A) $5

(B) $10

(C) $15

(D) $20

(E) The two seminars cost the same amount.

If you're not immediately sure how to proceed, jot down the numbers in an orderly fashion:

X	$700	20
Y	$750	15

This step only takes a moment and gets your brain moving. When you organize the information in this way, you may see that the next step involves division:

X $\quad \$700 \div 20 = \$35$

Y $\quad \$750 \div 15 = \$50$

Now you can easily see that Seminar Y cost $15 more than Seminar X, so the correct answer is Choice (C).

EXAMPLE

Jessica is in charge of stocking shelves at a supermarket. Today, she has already stocked 8 boxes that each contained 40 cans of soup, 12 boxes that each contained 24 cans of corned beef hash, and 4 boxes that each contained 60 cans of tuna. How many cans has Jessica stocked today?

(F) 148

(G) 310

(H) 624

(J) 848

(K) 1,020

Record the numbers in this question as follows:

8	40
12	24
4	60

Now multiply these numbers across (with or without your calculator, as needed) to get the number of cans in each set of boxes:

$$8 \times 40 = 320$$
$$12 \times 24 = 288$$
$$4 \times 60 = 240$$

Finish by adding the results: $320 + 288 + 240 = 848$. Therefore, the correct answer is Choice (J).

Sketching out problem information

Some word problems are much easier to solve when you draw a sketch to organize your thoughts. This technique is especially helpful if you're a visual learner. So if you like to draw, paint, or play video games, lead with your strength and try to find a visual way to express math problems whenever possible. The following example shows how to use a sketch to your advantage.

EXAMPLE

The 12:00 p.m. eastbound train left the station at a constant speed of 40 miles per hour. At 12:45 p.m., the next eastbound train left the station at a constant speed of 60 miles per hour. Assuming neither train stops along the way, how far apart will the two trains be at 2:00 p.m.?

(A) 5 miles

(B) 10 miles

(C) 12 miles

(D) 15 miles

(E) 18 miles

This problem is difficult to visualize, so sketching out the information can help you arrive at the correct answer:

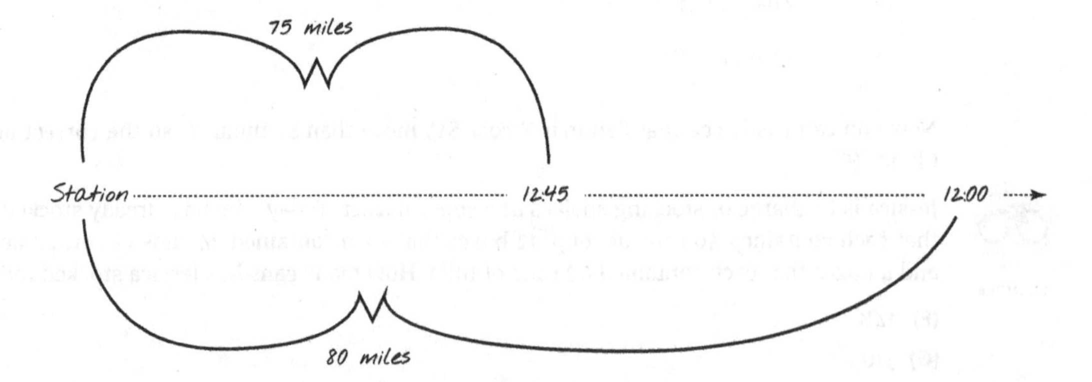

This figure helps illustrate a way into the problem: By 2:00 p.m., the 12:00 p.m. train has traveled for 2 hours at 40 miles per hour, so it's 80 miles from the station. And the 12:45 p.m. train has traveled for 1 hour and 15 minutes at 60 miles per hour. The 1 hour accounts for 60 miles. The 15 minutes is a quarter of an hour, so this accounts for 15 miles. Thus, the 12:45 p.m. train is 75 miles from the station. As a result, the trains are 5 miles apart, making the correct answer Choice (A).

2

Building Your Pre-Algebra and Elementary Algebra Skills

IN THIS PART . . .

Understanding factors and multiples, percents, ratios and proportions, and squares and square roots.

Working with algebraic expressions and solving algebraic equations.

Understanding the presentation of data in tables and graphs, calculating the mean and the median, and understanding simple, compound, and conditional probability.

Solving practice problems for all of these topics.

IN THIS CHAPTER

» Beefing up your arithmetic skills

» Working with factors and multiples

» Handling percents, ratios, and proportions

» Understanding powers and square roots

Chapter 4
Starting with the Basics: Pre-Algebra

Pre-algebra focuses on basic arithmetic skills, which are the basis for all of math. The good news is that you probably began learning arithmetic during your first few years in school, so nothing here will be new. Even so, beefing up your skills can help to improve your ACT score.

In this chapter, I start off with a review of the four basic operations — addition, subtraction, multiplication, and division. I focus on helping you answer ACT questions involving number sequences. I also review inequalities (such as <, >, ≥, and ≤) and absolute value. Next, you get some practice working with factors and multiples and a variety of questions that focus on percents, including percent increase and decrease. After that, I show you how to use ratios and proportions to answer ACT questions. To finish up, I provide information on powers and square roots.

Getting Back to Basics: Answering Basic Arithmetic Questions

Arithmetic is the basis for all of math, so you need a good grounding in it to do well on the math portion of the ACT. Fortunately, every question in this book requires a bit of arithmetic, so you get lots of practice. In this section, I start off with the basics of arithmetic, including the four basic operations, inequalities, and absolute value.

Reviewing the four basic operations

The four basic arithmetic operations, as you probably know, are addition, subtraction, multiplication, and division. You've been working with these operations for a long time, and you'll be using them a lot more on the ACT.

Many of the simpler word problems toward the beginning of the ACT give you scenarios that require you to apply basic arithmetic. If you read this type of question carefully, you should be able to work it out.

Arithmetic word problems aren't difficult, but they sometimes can be a little confusing. Use the room provided in your test booklet for scribbling notes as you work on them. Sometimes it even helps to draw a picture, as I discuss in Chapter 3.

Carla, a secretary in a law firm, is paid on a monthly basis. Each month, she works exactly 22 days and is paid $3,850. In May, Carla needed to take a day off. The firm deducted one day's pay from her paycheck and hired a temporary secretary named Jerome, paying him $135 for the day. How much did the firm pay out in total to both Carla and Jerome that month?

(A) $3,810

(B) $3,835

(C) $3,850

(D) $3,875

(E) $3,900

To answer this question, you first need to find out how much Carla is paid per day. To do this, divide her monthly salary ($3,850) by the number of days she works each month (22):

$$3,850 \div 22 = 175$$

As you can see, Carla is paid $175 dollars a day. When she took the day off, the firm paid Jerome $135, which is $40 less ($175 - 135 = 40$). Thus, the firm paid out $40 less to Carla and Jerome than it usually pays to Carla. Subtract that $40 from Carla's normal salary to find your answer:

$$3,850 - 40 = 3,810$$

Therefore, the firm paid $3,810, so the correct answer is Choice (A).

Jacqui works in a supermarket stocking shelves. She likes to time herself while she works and has found that it takes 10 seconds to open a box and 2 seconds to stock each item. About how long will it take her to open and stock a complete order of 24 boxes, each of which contains 36 cans of tuna?

(F) Between 1,200 and 1,400 seconds

(G) Between 1,400 and 1,600 seconds

(H) Between 1,600 and 1,800 seconds

(J) Between 1,800 and 2,000 seconds

(K) Between 2,000 and 2,200 seconds

The order contains $24 \times 36 = 864$ cans. Each can takes 2 seconds to stock, which in total takes $864 \times 2 = 1,728$ seconds. Additionally, opening all 24 boxes will take $24 \times 10 = 240$ seconds. Thus, the total order will take $1,728 + 240 = 1,968$ seconds. So the correct answer is Choice (J).

Following along with number sequences

A *number sequence* is a set of numbers that's in a specific order based on a rule. The rule that governs a sequence is usually an operation performed on each number to produce the next number in the sequence.

To answer most ACT questions involving a sequence, you need to discover the rule and then apply it. Begin by looking for a way to apply a basic operation to one number in the sequence in order to change it to the next number. Then, see if this rule works for the rest of the sequence.

TIP

When a sequence is increasing, look for ways to add or multiply (or both) one number to produce the next number.

EXAMPLE

In the number sequence 3, 10, 31, _____, 283, what number should be placed in the blank?

(A) 38

(B) 52

(C) 91

(D) 94

(E) 122

The rule to change 3 to 10 using only addition would be "add 7," because $3+7=10$. But this rule doesn't change 10 to 31, because $10+7=17$. Using only multiplication, the rule would be "multiply by $3\frac{1}{3}$," because $3\times 3\frac{1}{3}=10$. Again, however, this rule doesn't change 10 to 31, because $10\times 3\frac{1}{3}=33\frac{1}{3}$.

A rule involving multiplication and addition is "multiply by 3 and then add 1." This does the trick, because $10\times 3+1=31$. So here's how you find the next number in the sequence: $31\times 3+1=94$. Therefore, the correct answer is Choice (D).

In some cases, two different rules work to produce alternating numbers, as this next example illustrates.

EXAMPLE

If the first five numbers in a sequence are 5, 7, 14, 16, 32, what is the tenth number?

(F) 140

(G) 142

(H) 284

(J) 286

(K) 288

The two alternating rules producing this sequence are "add 2" and "multiply by 2." Use these rules to continue the sequence:

$$5, 7, 14, 16, 32, 34, 68, 70, 140, 142, \ldots$$

Thus, the correct answer is Choice (G).

Checking out inequalities

You probably know that an equation is a math statement that tells you two values are equivalent. In contrast, an *inequality* tells you that two values *aren't* equivalent. The four inequalities you'll see on the ACT are

- » Greater than (>)
- » Less than (<)
- » Greater than or equal to (≥)
- » Less than or equal to (≤)

REMEMBER

Note that > and < both tell you that two values are *not* equal, and ≥ and ≤ tell you that they *may* be equal.

TIP

Sometimes, reversing an *entire* inequality can help you make better sense of it without changing its meaning. For example, the inequality $2 > x$ may not be immediately clear, but its equivalent $x < 2$ is easier to understand: x is less than 2, so you can picture values that x may be (1, 0, −1, and so forth).

An ACT question involving an inequality may ask you to plot the *solution set* — the set of all possible solutions — on a number line. A black dot on the number line tells you that the value is included — that is, the sign is either ≥ or ≤. An open circle tells you that the value is *not* included — that is, the sign is either > or <.

EXAMPLE

Which of the following is the solution set for the inequality $-2 \le x < 3$

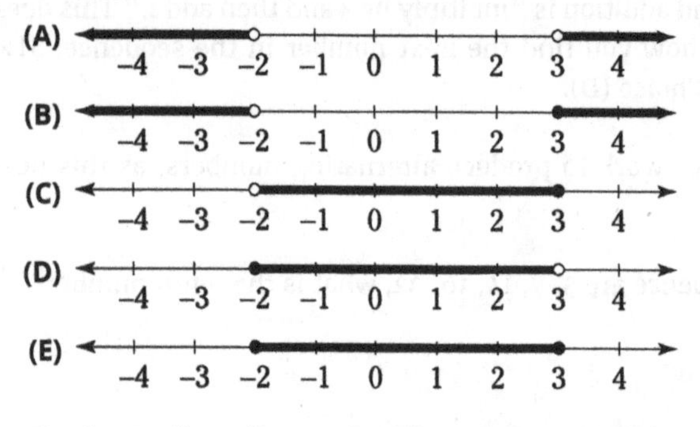

The inequality tells you that the solution set for x is between the two values, so you can rule out Choices (A) and (B). The value −2 is included, because the first inequality is ≤. So there should be a black dot at −2; thus, you can rule out Choice (C). The value 3 is *not* included, because the second inequality is <. As a result, there should be an open circle at 3; thus, you can rule out Choice (E). Therefore, the correct answer is Choice (D).

Evaluating a number's absolute value

The *absolute value* of a number is its value when you drop the minus sign. If the number doesn't have a minus sign to begin with, its value stays the same. For example:

$$|3| = 3 \qquad |-4| = 4 \qquad |0| = 0$$

TIP

Absolute value isn't difficult, but you can avoid confusion by taking ACT questions with absolute value in two steps: First, evaluate and remove the absolute value bars, and then complete the problem.

EXAMPLE

When working with absolute value, $|-4| - |-6| =$

(A) 10

(B) 2

(C) 0

(D) –2

(E) –10

Begin by evaluating $|-4|$ and $|-6|$ separately to remove the absolute value bars: $|-4| = 4$ and $|-6| = 6$. Thus, you can rewrite the equation:

$$|-4| - |-6| = 4 - 6$$

Now the problem becomes simple: $4 - 6 = -2$, so the correct answer is Choice (D).

In some cases, you may need to apply absolute value to a number line.

EXAMPLE

Which of the following number lines expresses the set of all possible values of x for the inequality $x \geq |-2|$?

(F)

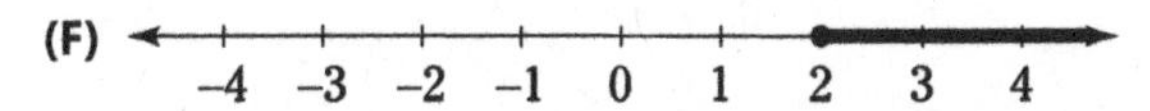

–4 –3 –2 –1 0 1 2 3 4

(G)

–4 –3 –2 –1 0 1 2 3 4

(H)

–4 –3 –2 –1 0 1 2 3 4

(J)

–4 –3 –2 –1 0 1 2 3 4

(K)

–4 –3 –2 –1 0 1 2 3 4

To begin, evaluate the absolute value: $|-2| = 2$. So you can rewrite the inequality as follows:

$$x \geq 2$$

As you can see, the solution set includes 2 and all values greater than 2, so the correct answer is Choice (F).

Understanding Factors and Multiples

Two important concepts in arithmetic are factors and multiples. Both of these are related to the simple idea of divisibility: One positive integer is *divisible* by another if you can divide the first integer by the second without leaving a remainder. For example:

- » 14 is divisible by 2, because $14 \div 2 = 7$ (with no remainder)
- » 14 is not divisible by 3, because $14 \div 3 = 4$ r 2 (4 with a remainder of 2)

When one number is divisible by another, you can describe the relationship between them using the words *factor* (the smaller number) and *multiple* (the larger number). For example:

- 2 is a factor of 14
- 14 is a multiple of 2

In this section, I show you how to answer a variety of ACT questions that involve factors and multiples. (For even more on these two topics, check out another book by Yours Truly: *Basic Math and Pre-Algebra For Dummies*, 2nd Edition [Wiley].)

Finding factors

The factors of a positive integer are always less than the number itself, so it's easy to list all the factors of a number. To answer an ACT question, you may need to generate a list of all factors of a number.

EXAMPLE

Which of the following integers has exactly the same number of factors as the number 21?

(A) 3

(B) 4

(C) 5

(D) 6

(E) 7

Of course, to figure out this problem you have to know how many factors 21 has. Every integer greater than 1 has at least two factors: 1 and the number itself. To list all the factors of a number, begin by writing down these two numbers with some space between them, like this:

Factors of 21: 1 21

All other factors must fall between 1 and 21. Start by testing 2: Clearly 21 isn't divisible by 2, because 21 isn't even. Next, test whether 21 is divisible by 3: Yes, because $21 \div 3 = 7$ with no remainder. So you can add 3 to the early part of the list and 7 to the later part:

Factors of 21: 1 3 7 21

Now all other factors (if any exist) must fall between 3 and 7. A quick test of 4, 5, and 6 shows that 21 isn't divisible by any of these numbers. So 21 has exactly four factors: 1, 3, 7, and 21.

Use the same process to list the factors of the five answer choices until you find one that has exactly four factors:

Factors of 3:	1	3		
Factors of 4:	1	2	4	
Factors of 5:	1	5		
Factors of 6:	1	2	3	6

As you can see, 21 and 6 both have exactly four factors, so the correct answer is Choice (D).

Knowing how to list the factors of a number in this way also is important for finding the *greatest common factor* (GCF) of a set of numbers — that is, the highest number that's a factor of every number in that set.

EXAMPLE

What is the greatest common factor of 24, 36, and 50?

(F) 2

(G) 3

(H) 4

(J) 6

(K) 8

To answer this question, list the factors of all three numbers:

Factors of 24:	1	2	3	4	6	8	12	24	
Factors of 36:	1	2	3	4	6	9	12	18	36
Factors of 50:	1	2	5	10	25	50			

The greatest number that appears in all three lists is 2, so the correct answer is Choice (F).

Generating lists of multiples

The positive multiples of an integer are always greater than the integer itself. For this reason, you can never list all the multiples of a number (as you can with factors; see the preceding section for details). When answering an ACT question, it's often handy to make a quick list of the first ten or so multiples of a number. Making this list is simple. Just keep adding the number you're working with. For example:

Multiples of 4: 4, 8, 12, 16, 20, 24, 28, 32, 36, 40 ...

EXAMPLE

When adding the fractions $\frac{2}{3}, \frac{3}{5}, \frac{7}{10}$, and $\frac{14}{15}$, what is the lowest common denominator?

(A) 15

(B) 20

(C) 30

(D) 45

(E) 60

The denominators are the bottom numbers in the fractions: 3, 5, 10, and 15. To find the lowest common denominator among these four numbers, generate a list of the multiples of these numbers until you find the lowest number that appears on all three lists, like so:

Multiples of 3: 3, 6, 9, 12, 15, 18, 21, 24, 27, **30**
Multiples of 5: 5, 10, 15, 20, 25, **30**, 35, 40
Multiples of 10: 10, 20, **30**, 40, 50, 60
Multiples of 15: 15, **30**, 45, 60

As you can see, 30 is the lowest common denominator for these four fractions, so the correct answer is Choice (C).

Perfecting Your Knowledge of Percents

To do well on the ACT, you need to have a solid understanding of percents. Questions may ask you to calculate a percentage directly. Alternatively, you may need to use percents to handle a word problem or work with a graph, such as a pie chart.

Before you begin reviewing percentage problems, make sure your calculator has a percent key to help you save time when calculating percents. Even with a calculator to help, however, certain percent problems can be confusing. So read on for details.

In this section, I show you a variety of ways to answer percent questions. I also give you some practice solving the much-dreaded percent increase and percent decrease problems.

Knowing basic percent conversions

TIP

Even though you can use a calculator on the ACT, you can save a lot of time by knowing a few basics. One good place to start is by memorizing the basic conversions for percents, decimals, and fractions. I list the most common conversions in Table 4-1.

TABLE 4-1 Converting Basic Percents to Decimals and Fractions

Percent	Decimal	Fraction
1%	0.01	$\frac{1}{100}$
10%	0.1	$\frac{1}{10}$
20%	0.2	$\frac{1}{5}$
25%	0.25	$\frac{1}{4}$
30%	0.3	$\frac{3}{10}$
$33\frac{1}{3}\%$	0.3...	$\frac{1}{3}$
40%	0.4	$\frac{2}{5}$
50%	0.5	$\frac{1}{2}$
60%	0.6	$\frac{3}{5}$
$66\frac{2}{3}\%$	0.6...	$\frac{2}{3}$
70%	0.7	$\frac{7}{10}$
75%	0.75	$\frac{3}{4}$
80%	0.8	$\frac{4}{5}$
90%	0.9	$\frac{9}{10}$
100%	1	1

Taking advantage of a quick trick for calculating some percents

Ready for a way to save a few seconds on the ACT? In this section, I let you in on a quick trick for finding certain percents. You simply have to reverse the numbers you're working with. Check out the following example.

If 80% of 50 is x, what is x percent of 25?

(A) 10

(B) 20

(C) 40

(D) 80

(E) 100

Of course, you can do 80% of 50 on your calculator, but an easier way in this case is to reverse the two numbers:

$$80\% \text{ of } 50 = 50\% \text{ of } 80$$

This trick always works. In this case, you can see that 50% of 80 equals half of 80, which is 40. So $x = 40$. Now you have to find 40% of 25, so use the trick again:

$$40\% \text{ of } 25 = 25\% \text{ of } 40$$

Again, you probably can see that 25% of 40 equals one-quarter of 40, which is 10. Therefore, the correct answer is Choice (A).

Calculating straightforward percent problems

Even if you're really good at working with fractions, decimals, and percents, you can save yourself a lot of trouble by using your calculator to work out hairy calculations.

First of all, be sure that your calculator has a percent key. Sure, technically you can calculate percents by converting them to decimals — and as a math guy, it's my duty to recommend that you know how to do this! But on the ACT, you don't need to waste the time and risk making an error. So take full advantage of your calculator. The following example shows you how.

When you calculate 62.5% of 450, how many decimal places does the resulting number have?

(A) 0

(B) 1

(C) 2

(D) 3

(E) 4

You can answer this question by answering 0.625×450. But when rushing on the test, if you convert the decimal incorrectly, you'll get the wrong answer. For this problem, your calculator is your best friend:

$$62.5\% \times 450 = 281.25$$

The resulting number has two decimal places, so the correct answer is Choice (C).

Untangling complex percent problems

Using your calculator to find a result is straightforward when you know the number you're calculating with and the percentage you're taking. For example, you may need to determine what 50% of 10 is. With or without a calculator, you can find out pretty quickly that the answer to this question is 5. But for some other ACT questions, you're given information that's more difficult to untangle.

TIP

When answering this type of question, the trick is to translate it into an equation. Substitute either 0.01 for the percent sign, and let x (or any variable you like) stand for the number you're trying to find. Then use algebra to solve for x (flip to Chapter 5 for a review of elementary algebra).

In this section, I provide two examples as practice for the problems you may see on the ACT.

EXAMPLE

What percent of 600 is 270?

(A) $37\frac{1}{2}\%$

(B) 40%

(C) 45%

(D) $222\frac{2}{9}\%$

(E) 250%

A good way to solve this type of percent problem is to translate the words into an equation. (I discuss this technique further in Chapter 5.) Here's how the example problem would look:

$$(x\%)(600)=270$$

Substitute 0.01 for the percent sign in this equation and solve for x:

$$\begin{aligned} x(0.01)(600) &= 270 \\ 6x &= 270 \\ x &= 45 \end{aligned}$$

Thus, 45% of 600 is 270, so the correct answer is Choice (C).

EXAMPLE

If 40% of a is 30, then a is what percent of 50?

(F) 33%

(G) $66\frac{2}{3}\%$

(H) 75%

(J) $133\frac{1}{3}\%$

(K) 150%

To answer this question, you first need to find the value of a, so make an equation from the values in the first part of the question. Here's how it would look:

$$(40\%)(a)=30$$
$$40(0.01)(a)=30$$

Now solve this equation for *a*:

$$0.4a = 30$$
$$a = \frac{30}{0.4}$$
$$a = 75$$

Thus, $a = 75$. Finally, you make an equation from the values in the second part of the question. Here's how: "75 is what percent of 50" becomes

$$75 = (x\%)(50)$$

Substitute 0.01 for the percent sign in this equation and solve for *x*:

$$75 = x(0.01)(50)$$
$$75 = 0.5x$$
$$\frac{75}{0.5} = x$$
$$150 = x$$

So the correct answer is Choice (K).

Handling the ups and downs of percent increase and decrease

Many students find questions that involve percent increase and percent decrease somewhat confusing. The first step to answering these questions correctly is identifying them when they're presented. Some common scenarios for percent increase questions include the following:

- Sales tax added to the price of an item
- Tipping a server at a restaurant
- Interest earned on an investment

Some typical situations for percent decrease questions are

- Money lost on an investment
- Discount on an item being sold
- Deduction from a paycheck due to taxes

After you know whether you're dealing with percent increase or percent decrease, here's how you handle the calculations:

- **Increase:** Calculate a percent increase as $100\% +$ the percent. For example, a percent increase of 15% is equal to $100\% + 15\% = 115\%$ of the original value.
- **Decrease:** Calculate a percent decrease as $100\% -$ the percent. For example, calculate a percent decrease of 20% as $100\% - 20\% = 80\%$.

The following two examples show you how to handle both types of questions from start to finish.

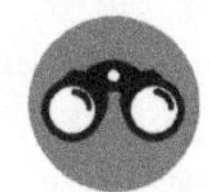

EXAMPLE

Randy bought a small guitar amplifier priced at $165 with a special coupon that gave him a 15% discount. About how much did he end up paying for the amp?

(A) Less than $100

(B) Between $100 and $120

(C) Between $120 and $140

(D) Between $140 and $160

(E) More than $160

A 15% discount is a percent decrease of 15%:

$$100\% - 15\% = 85\%$$

Use your calculator to find this percentage of the original price of $165:

$$85\% \times \$165 = \$140.25$$

Thus, the correct answer is Choice (D).

You can apply this method for finding percent increase and decrease to any of the percent problems that I discuss earlier in the chapter, in "Untangling complex percent problems."

EXAMPLE

Keith's portfolio is currently worth $10,200, representing a 20% increase on his original investment. How much did he originally invest?

(F) $7,800

(G) $8,160

(H) $8,440

(J) $8,500

(K) $8,880

A 20% increase is calculated as $100\% + 20\% = 120\%$, so use the following formula:

$$(120\%)(n) = 10{,}200$$

Change the percent sign to 0.01 and solve:

$$\begin{aligned} 120(0.01)(n) &= 10{,}200 \\ (1.2)n &= 10{,}200 \\ n &= \frac{10{,}200}{1.2} \\ n &= 8{,}500 \end{aligned}$$

Therefore, the correct answer is Choice (J).

Using Ratios and Proportions to Compare Quantities

A *ratio* is a comparison of two quantities based on the operation of division. For example, if a school has one teacher for every eight students, you can express the teacher-to-student ratio in any of the following ways:

$1:8$ $\qquad$ 1 to 8 $\qquad$ $\frac{1}{8}$

Notice that this ratio expresses the ratio of *teachers* to *students*. Thus, the 1 goes before the 8 and, in the fraction, the 1 goes on top of the 8.

When answering an ACT question that includes a ratio, a good strategy is to express the ratio as an equivalent fraction. Then you can pull out all the tools you already have for working with fractions — for example, reducing, converting to decimals, and so forth.

A company has a total of 150 employees, 25 of whom are managers. What is the ratio of managers to non-managers?

(A) 1 to 3

(B) 1 to 4

(C) 1 to 5

(D) 1 to 6

(E) 2 to 5

The company has 25 managers, so the remaining 125 employees are non-managers. Express this ratio as a fraction and then reduce it:

$$\frac{\text{Managers}}{\text{Non}-\text{managers}} = \frac{25}{125} = \frac{1}{5}$$

The ratio of managers to non-managers is 1 to 5, so the correct answer is Choice (C).

One of the most practical applications of the ratio is a *proportion*, which is an equation based on a ratio. For example, if you know the ratio of boys to girls, you can express this as a fraction, set it equal to another fraction that includes a variable, and then solve. The following example illustrates how this concept works.

A summer camp has a boy-to-girl ratio of 8:11. If the camp has 88 boys, what is the total number of children at the camp?

(F) 121

(G) 128

(H) 152

(J) 176

(K) 209

Begin by setting up the proportion as the following equation:

$$\frac{\text{Boys}}{\text{Girls}} = \frac{8}{11}$$

Before continuing, notice that the ratio specifically mentions boys first and girls second, so this order is maintained in the equation. The camp has 88 boys, so substitute this number for *Boys* in the equation. You don't know how many girls there are, so use the variable *g*. Here's what your equation now looks like:

$$\frac{88}{g} = \frac{8}{11}$$

To find out how many girls are at the camp, solve for *g* using algebra, which I discuss in greater detail in Chapter 5. First, cross-multiply to get rid of the two fractions:

$$\begin{aligned} 88(11) &= 8(g) \\ 968 &= 8g \end{aligned}$$

Now divide both sides by 8:

$$121 = g$$

The camp includes 121 girls and 88 boys, so you know it has a total of 209 children; therefore, the correct answer is Choice (K).

Working with Powers and Square Roots

When you raise a number to a *power* (also called an *exponent*), you multiply that number by itself a given number of times. Powers and *roots* — the inverse operation to powers — are important operations to understand for the ACT. In this section, I walk you through a variety of practice problems using these skills.

Harnessing the power of exponents

When you raise a number to a *power*, you multiply that number by itself repeatedly. For example:

$$3^4 = 3 \times 3 \times 3 \times 3 = 81$$

Raising a number to a power involves two numbers: The *base* is the number that's multiplied and the *exponent* is the number of times that it's multiplied. You can use any number as a base, including negative numbers and fractions. For example:

$$(-2)^5 = (-2)(-2)(-2)(-2)(-2) = -32 \qquad \left(\frac{4}{5}\right)^3 = \left(\frac{4}{5}\right)\left(\frac{4}{5}\right)\left(\frac{4}{5}\right) = \frac{64}{125}$$

A minus sign in the exponent changes the result to the *reciprocal* of the result — that is, the numerator (top number) and denominator (bottom number) are reversed. Or, if a number has no denominator, the numerator becomes the denominator. Consider these examples:

$$3^{-4} = \frac{1}{81} \qquad -2^{-5} = -\frac{1}{32} \qquad \left(\frac{4}{5}\right)^{-3} = \frac{125}{64}$$

When you raise a number to a fractional exponent, take the numerator as a power and the denominator as a square root. The most common fractional exponent is $\frac{1}{2}$, which equals a square root. For example:

$$25^{\frac{1}{2}} = \sqrt{25} = 5$$

EXAMPLE

If $\left(\frac{3}{4}\right)^2 < k < 16^{\frac{1}{2}}$, how many different integers could k equal?

(A) 0

(B) 1

(C) 2

(D) 3

(E) More than 3

To begin, evaluate the two powers:

$$\left(\frac{3}{4}\right)^2 = \frac{9}{16} \qquad 16^{\frac{1}{2}} = \sqrt{16} = 4$$

You can rewrite the inequality as follows:

$$\frac{9}{16} < k < 4$$

Therefore, k could equal 1, 2, or 3, so the correct answer is Choice (D).

Squaring off: Squares and square roots

Squares and square roots are common on the ACT. In this section, I show you how to work with them to answer basic ACT questions.

Squares

When you *square* a number, you multiply it by itself. The symbol that denotes a square number is a superscript 2. For example:

$$4^2 = 4 \times 4 = 16$$

It's a good idea to be familiar with the square numbers up to 100. Table 4-2 can help you bone up on them.

TABLE 4-2 The Square Numbers from 1^2 to 10^2

1^2	2^2	3^2	4^2	5^2	6^2	7^2	8^2	9^2	10^2
1	4	9	16	25	36	49	64	81	100

EXAMPLE

The product of two consecutive square numbers is 900. What is the sum of these two numbers?

(A) 25

(B) 41

(C) 61

(D) 85

(E) 113

A quick comparison of values for the first ten square numbers shows you that the pair of numbers is 25 and 36, because $25 \times 36 = 900$. Their sum is $25 + 36 = 61$, so the correct answer is Choice (C).

Square roots

The inverse operation to squaring is taking a *square root* (also called a *radical*). When you take a square root of a number, you find a value that, when multiplied by itself, produces the number you started with. For example:

$$\sqrt{16} = 4 \textit{ because } 4 \times 4 = 16$$

Taking the square root of a negative number produces an *imaginary number,* which is outside the set of real numbers.

Although ACT questions may include imaginary numbers, they're rare. Your ACT probably will have no more than one question that includes imaginary numbers. (For more on working with imaginary numbers, see Chapter 12.)

If x is divisible by 4 and $0 < x < \sqrt{49}$, what is the value of x^2?

(A) 4

(B) 8

(C) 16

(D) 32

(E) 36

The value of $\sqrt{49}$ is 7, so:

$$0 < x < 7$$

The variable x is divisible by 4, and 4 is the only number between 0 and 7 that's divisible by 4, so $x = 4$. Thus, $x^2 = 4^2 = 16$, so the correct answer is Choice (C).

IN THIS CHAPTER

- » **Knowing two main ideas and six key words of algebra**
- » **Evaluating, simplifying, and factoring expressions**
- » **Solving a wide variety of equations**
- » **Turning words into an equation you know how to solve**

Chapter 5 It's Elementary Algebra, My Dear Watson

Algebra is a key focus of ACT math. Luckily, in this chapter and in Chapter 8, you gain the algebra skills you need most to ace the ACT. I start at the very beginning, reviewing the basic concepts and vocabulary of algebra so you gain a good foundation for what follows. Then I show you three key skills that are essential for working with algebraic expressions: evaluating, simplifying, and factoring. At that point, you're ready to solve a variety of equations that you'll see more of on the ACT: rational equations, square root (radical) equations, absolute value equations, and exponential equations. I also show you how to handle equations that have more than one variable. To wrap up, I cover some basics of word problems, showing you how to do direct translation of words into equations.

Knowing the Two Big Ideas of Algebra

At its core, solving an algebra equation comes down to two big ideas — the *how* and the *what* of algebra:

- » **How algebra works:** Every step you take must keep the equation in balance.
- » **What you need to do:** Isolate the variable on one side of the equal sign.

These two main points may seem a little scary, but don't worry. In this section, I discuss everything you need to know to easily solve any algebra equation that the ACT throws at you.

Every step you take: Keeping equations balanced

You can change an equation in all sorts of ways, provided every step you take keeps it in balance. When I say you have to keep an equation balanced, what I mean is that you need to make the same changes to both sides of the equation.

For example, here's an equation you know is true:

$$2+2=4$$

You can add the same number to or subtract the same number from both sides of this equation. For example:

$$2+2+3=4+3$$

The result here is still true because both sides equal 7. Similarly, you can multiply or divide both sides by the same number. For example:

$$5(2+2)=5(4)$$

Again, the equation stays in balance because both sides equal 20. You can even raise both sides of the equation to a power and still keep it in balance. For example:

$$(2+2)^2=4^2$$

Both sides of the equation equal 16, so the equation is still in balance.

Walkin' a lonely road: Isolating the variable

Your main objective in algebra is to *isolate the variable* — that is, you want to get the variable alone on one side of the equation and everything else on the other side. The *variable* is simply a letter that stands in for a specific number. The most common variable is x, but you can use any letter you want.

EXAMPLE

What is the value of x if $3x+4=12-x$?

(A) 2

(B) 4

(C) 6

(D) 8

(E) 10

To answer this question, you want to get all the x's on one side of the equation and all the numbers on the other side. To do so, begin by adding x to both sides:

$$\begin{array}{l} 3x+4=12-x \\ \underline{+x \qquad\quad +x} \\ 4x+4=12 \end{array}$$

This step brings all the x's to the left side of the equation. Next, subtract 4 from both sides:

$$\begin{array}{l} 4x+4=12 \\ \underline{\quad -4\ -4} \\ 4x \quad\ =8 \end{array}$$

To finish, divide both sides by 4:

$$\frac{4x}{4} = \frac{8}{4}$$
$$x = 2$$

Therefore, $x = 2$, so the correct answer is Choice (A).

REMEMBER

Although most of the algebra questions you'll see on the ACT will be more difficult than this one, every algebra equation comes back to this simple strategy of isolating the variable.

Discovering Six Choice Words about Algebra

Algebra is chock-full of words that are useful but often misunderstood. Here are a few favorites:

- **Variable:** A *variable* is any letter that stands for a number. The most commonly used letters are *x* and *y*, but you can use any letter.
- **Constant:** A *constant* is a number without a variable. For example: The equation $6m + 7 = -m$ has one constant, the number 7.
- **Equation:** An *equation* is any string of numbers and symbols that makes sense and includes an equal sign. For example, here are three equations:

  $2 + 2 = 4$ $\qquad 3x - 7 = 46 \qquad \left(\frac{1}{25}\right)^{n+1} = \sqrt{5}$

- **Expression:** An *expression* is any string of numbers and symbols that makes sense when placed on one side of an equation. For example, here are four expressions:

  $2 + 2 \qquad 3x - 7 \qquad -x^2 - 9x + 11 \qquad \left(\frac{1}{25}\right)^{n+1}$

- **Term:** A *term* is any part of an expression that's separated from the other parts by either a plus sign (+) or a minus sign (-). ***Important:*** A term always includes the sign that immediately precedes it. For example:

  The expression $3x - 7$ has two terms: $3x$ and -7.

  The expression $-x^2 - 9x + 11$ has three terms: $-x^2$, $-9x$, and 11.

  The expression $\left(\frac{1}{25}\right)^{n+1}$ has one term: $\left(\frac{1}{25}\right)^{n+1}$.

- **Coefficient:** A *coefficient* is the numerical part of a term, including the sign that precedes it (+ or -). ***Important:*** Every term has a coefficient. When a term appears to have no coefficient, its coefficient is either 1 or –1, depending on the sign. For example:

  The term $3x$ has a coefficient of 3.

  The term $-9x$ has a coefficient of –9.

  The term $-x^2$ has a coefficient of –1.

Being clear about the meanings of these six words can help you with the rest of this chapter and with any other math you study.

Express Yourself: Working with Algebraic Expressions

An *expression* is any string of numbers and symbols that makes sense when placed on one side of an equation, as I discuss in the earlier section "Discovering Six Choice Words about Algebra." In this section, I show you three important skills for working with expressions:

- **Evaluating:** Plugging in numbers for variables to find the value of the expression
- **Simplifying:** Reducing the complexity of an expression
- **Factoring:** Pulling a common factor out of each term of an expression

REMEMBER

Working with expressions is important for solving equations. Additionally, some ACT questions may test these skills directly.

Evaluating expressions

When you *evaluate* an expression, you find out what number it's equal to. To evaluate an algebraic expression — that is, an expression with at least one variable — you need to know the values of all the variables in that expression. Then you can *substitute*, or plug in, the value of each variable and use the normal order of operations (PEMDAS) to find the value of the expression. For more on the order of operations, check out Chapter 3.

REMEMBER

Make sure you know how to evaluate expressions correctly. If you need more review on this topic, I cover it in much greater detail in *Basic Math and Pre-Algebra For Dummies*, 2nd Edition (Wiley).

EXAMPLE

What is the value of $3a^2 - ab - 2b^2$ if $a = 4$ and $b = -5$?

(A) 10

(B) 16

(C) 18

(D) 20

(E) 25

To begin, substitute 4 for a and –5 for b throughout the expression:

$$3a^2 - ab - 2b^2 = 3(4)^2 - 4(-5) - 2(-5)^2$$

Now evaluate the expression using the standard order of operations. First, evaluate the powers:

$$= 3(16) - 4(-5) - 2(25)$$

Next, multiply:

$$= 48 + 20 - 50$$

Finally, add and subtract:

$$= 18$$

Therefore, the correct answer is Choice (C).

Simplifying expressions

When you *simplify* an expression, you make it more compact and easier to work with without changing its value. In this section, I show you a few basic techniques for simplifying expressions.

Combining like terms

Two terms in an expression are *like terms* when they have exactly the same variables, complete with the exactly the same exponents on each variable. Here are three examples of like terms:

$3x$ and $4x$ $\qquad$ $10k^2$ and $-9k^2$ $\qquad$ $2a^3b^2c$ and $-2a^3b^2c$

When you *combine like terms*, you put the terms together by adding their coefficients and keeping the same variable. For example:

$3x + 4x = 7x$ $\qquad$ $10k^2 + (-9k^2) = k^2$ $\qquad$ $2a^3b^2c + (-2a^3b^2c) = 0$

As you can see in the last case, when the coefficients of a pair of like terms add up to 0, the two terms cancel each other out.

EXAMPLE

Which of the following expressions is equivalent to $2x - 3y + 7xy - y - 2x$?

(A) $-2xy$

(B) $-3y + 7xy$

(C) $-4y + 7xy$

(D) $4x + 4y + 7xy$

(E) $4x - 4y + 7xy$

The expression contains two x terms that cancel each out, because $2x + (-2x) = 0$:

$$2x - 3y + 7xy - y - 2x = -3y + 7xy - y$$

It also contains two y terms, which combine as $-3y + (-y) = -4y$, leaving you with

$$= -4y + 7xy$$

Thus, the correct answer is Choice (C).

Removing parentheses

Some expressions contain parentheses that prevent you from combining like terms (as I describe in the preceding section). However, after you remove the parentheses from an expression, you may find that you can simplify the expression further. Generally speaking, three different cases require you to remove a single set of parentheses. Here are the rules for all three cases:

- **Parentheses preceded by a plus sign (+):** Drop the parentheses. For example:

  $$2x + (3 - 4y) = 2x + 3 - 4y$$

- **Parentheses preceded by a minus sign (–):** Negate — that is, change the sign of — every term inside the parentheses, and then drop the parentheses. For example:

  $$2x - (3 - 4y) = 2x - 3 + 4y$$

» **Parentheses preceded by a term outside the parentheses:** Distribute the term outside the parentheses — that is, multiply it by every term inside the parentheses — and then drop the parentheses. For example:

$2x(3-4y)=6x-8xy$

REMEMBER

When simplifying an expression with *nested parentheses* — that is, one set of parentheses inside another — always begin with the inside set and move outward.

EXAMPLE

Which of the following is equivalent to $4x^2-(3-2x(5-x))$?

(A) $x^2+10x+3$

(B) x^2-13x

(C) $2x^2+10x-3$

(D) $2x^2-7x$

(E) $6x^2-13x-3$

To simplify this expression, begin with the inside set of parentheses. The term $-2x$ precedes the parentheses, so distribute this term (multiply it by 5 and then by $-x$), and then drop this set of parentheses:

$$4x^2-(3-2x(5-x))=4x^2-\left(3-10x+2x^2\right)$$

Now a minus sign precedes the remaining set of parentheses, so negate each term (change its sign) and drop the parentheses:

$$=4x^2-3+10x-2x^2$$

To finish, combine the two like terms ($4x^2$ and $-2x^2$) and rearrange the terms in descending order by their exponents:

$$=2x^2-3+10x$$
$$=2x^2+10x-3$$

So the correct answer is Choice (C).

Getting FOILed

When simplifying two sets of adjacent parentheses, multiply every term in one set of parentheses by every term in the other. This process often is called *FOILing*, an acronym that stands for *First, Outside, Inside, Last.* The following example shows you how to FOIL an expression.

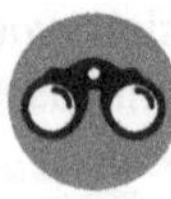

EXAMPLE

Which of the following is equivalent to the expression $(2x+3)(4x-1)$?

(A) $6x-3$

(B) $6x^2-10x-3$

(C) $8x^2+10x+3$

(D) $8x^2+10x-3$

(E) $8x^2+14x-3$

The acronym *FOIL* helps you keep track of the four multiplication calculations you need to perform in order to drop the parentheses. Multiply the two *first* terms, the two *outside* terms, the two *inside* terms, and finally the two *last* terms:

$$\textbf{First}: (2x)(4x) = 8x^2$$
$$\textbf{Outside}: (2x)(-1) = -2x$$
$$\textbf{Inside}: (3)(4x) = 12x$$
$$\textbf{Last}: (3)(-1) = -3$$

Place these four terms in one expression:

$$8x^2 - 2x + 12x - 3$$

Combine like terms:

$$8x^2 + 10x - 3$$

So the correct answer is Choice (D).

Simplifying expressions with exponents

When an exponent appears outside a set of parentheses, multiply everything in the parentheses by itself the number of times indicated. Often, doing so requires you to FOIL the contents of the parentheses as I show you in the previous section.

EXAMPLE

When simplified, how many terms are in the expression $(x+1)^3$?

(A) 1

(B) 2

(C) 3

(D) 4

(E) More than 4

To begin, change the power to multiplication:

$$(x+1)^3 = (x+1)(x+1)(x+1)$$

Now FOIL the contents of the first two parentheses. I do this in two steps:

$$= \left(x^2 + x + x + 1\right)(x+1)$$
$$= \left(x^2 + 2x + 1\right)(x+1)$$

Next, distribute the contents of the parentheses. To do this, multiply each of the three terms inside the first set of parentheses by both terms inside the second set — six multiplications in all:

$$= x^3 + x^2 + 2x^2 + 2x + x + 1$$

To finish, combine both sets of like terms:

$$= x^3 + 3x^2 + 3x + 1$$

The simplified form has four terms, so the correct answer is Choice (D).

Factoring expressions

Factoring an expression reverses the simplification process and introduces parentheses into the expression. Some equations are much easier to solve after you do some factoring. In this section, I show you two of the most useful ways to factor to solve elementary algebra problems. (By the way, I save factoring quadratic expressions for Chapter 8, where you need this skill for solving quadratic equations.)

Finding common factors

The most straightforward way to factor an expression is to look for the *greatest common factor* (*GCF*) among all the terms. You probably already have experience at finding the GCF of two or more integers (if not, check out Chapter 4). When finding the GCF among a set of terms, you use this skill and add in one more step: You identify the lowest exponent for each individual variable.

EXAMPLE

What is the greatest common factor among $6a^2b^5c^4, 12a^3b^3c^3$, and $15a^4b^8c$?

(A) $2a^2b^3c^4$

(B) $3a^2b^3c$

(C) $3a^2b^5c$

(D) $3a^4b^8c^4$

(E) $6a^2b^3c^3$

To begin, find the GCF of the three coefficients — 6, 12, and 15. In other words, find the highest integer that divides all three of them. The GCF of these three numbers is 3, so 3 must be the coefficient of the GCF; therefore, you can rule out Choices (A) and (E).

Next, focus on the variable a and find the *lowest* exponent among all the a's in the three terms. The exponents of a are 2, 3, and 4, so the lowest exponent is 2; therefore, the GCF you're looking for must have the variable a^2, so you can rule out Choice (D). Now focus on the variable b and find the lowest exponent among all the b's. The exponents of b are 5, 3, and 8, so the lowest is 3; therefore, the GCF has the variable b^3, so you can rule out Choice (C). Therefore, the correct answer is Choice (B).

Unearthing three useful ways to factor squares and cubes

Three handy formulas exist for factoring different combinations of squared and cubed variables. Most important is the formula for the *difference of two squares*, which looks like this:

$$x^2 - y^2 = (x + y)(x - y)$$

This formula comes in handy for solving some algebra problems that may otherwise force you to solve a system of equations (as I discuss in Chapter 8). I illustrate this in the following example.

EXAMPLE

If $x^2 - y^2 = 3$ and $x - y = 6$, what is the value of $x + y$?

(A) 1

(B) 2

(C) 3

(D) $\frac{1}{2}$

(E) $\frac{1}{3}$

TIP

You could solve this problem as a system of equations, first solving for x and then for y and then adding the results. But this would be time-consuming and tedious. Instead, noticing that the first equation contains the difference of two squares gives you a much simpler way to go.

Begin with the formula for the difference of two squares:

$$x^2 - y^2 = (x+y)(x-y)$$

Plug in 3 for $x^2 - y^2$ and 6 for $(x-y)$:

$$3 = (x+y)(6)$$

You want to isolate the expression $x + y$, so divide both sides by 6 and simplify:

$$\frac{3}{6} = x + y$$

$$\frac{1}{2} = x + y$$

Therefore, the correct answer is Choice (D).

Also useful are the formulas for the *sum of two cubes* and the *difference of two cubes:*

$$x^3 + y^3 = (x+y)\left(x^2 - xy + y^2\right) \qquad x^3 - y^3 = (x-y)\left(x^2 + xy + y^2\right)$$

As with the formula for the difference of squares, when these formulas are useful, the problems almost scream for you to use them!

EXAMPLE

Suppose that $x + y = a$, $x^2 + y^2 = b$, and $x^3 + y^3 = c$. What is the value of xy in terms of a, b, and c?

(F) $\frac{ab+c}{a}$

(G) $\frac{ab-c}{a}$

(H) $\frac{c-ab}{a}$

(J) $\frac{c-ab}{b}$

(K) $\frac{c-ab}{c}$

This problem threatens to be a math nightmare. But the right approach turns it into, well, merely a slightly unpleasant dream. Begin with the formula for the sum of two cubes:

$$x^3 + y^3 = (x+y)\left(x^2 - xy + y^2\right)$$

Rearrange the terms inside the second set of parentheses:

$$x^3 + y^3 = (x+y)\left(x^2 + y^2 - xy\right)$$

Now plug in a for $x + y$, b for $x^2 + y^2$, and c for $x^3 + y^3$:

$$c = a(b - xy)$$

Now the trick is to isolate xy:

$$\begin{aligned} c &= ab - axy \\ c - ab &= -axy \\ \frac{c-ab}{-a} &= xy \end{aligned}$$

This solution doesn't quite match any of the answers provided, so multiply the numerator and the denominator of the fraction by –1 and rearrange the variables in the top of the fraction alphabetically:

$$\frac{ab-c}{a}=xy$$

Thus, the correct answer is Choice (G). This problem is by no means easy, but knowing the right formula makes it solvable using the algebra you already know.

Solving Everything but the Kitchen Sink

Although the main ideas behind algebra are simple (as I discuss in the earlier section "Knowing the Two Big Ideas of Algebra"), executing them isn't always a walk in the park. Each type of ACT question requires a different strategy to find the correct answer quickly. So to help you get a firm grasp on them, in this section I show you how to answer a bunch of different elementary algebra questions.

Approaching rational equations rationally

At some point on the ACT, you'll likely come across a rational equation question. A *rational equation* includes one or more fractions with variables. A good first step with rational equations is to eliminate the fractions, turning the equation into something more friendly. When both sides of an equation are entirely fractions, the quickest way to eliminate the fractions is to *cross-multiply* — multiply the numerator of each fraction by the denominator of the other fraction and set the two products equal.

EXAMPLE

If $\frac{21n-6}{7}=\frac{4n+1}{3}$, then $7n=$

(A) 1

(B) 2

(C) 3

(D) 4

(E) 5

To cross-multiply, multiply $21n - 6$ by 3 and $4n+1$ by 7, and then set the two values equal to each other:

$$3(21n-6)=7(4n+1)$$

Now distribute both sides of the equation and solve for n:

$$63n-18=28n+7$$
$$35n-18=7$$
$$35n=25$$

Divide both sides by 5:

$$7n=5$$

This equation answers the question that's being asked, so you don't have to solve further; the correct answer is Choice (E).

For a rational equation with more than one fraction, you may need stronger medicine. One approach is to find a common denominator for all the terms in the equation and multiply every term by this number, like I show you in the following example.

EXAMPLE

If $\frac{3x}{4}+\frac{7}{6}=\frac{5x}{2}$, then $x=$

(F) $\frac{1}{2}$

(G) $\frac{1}{3}$

(H) $\frac{2}{3}$

(J) $\frac{1}{4}$

(K) $\frac{3}{4}$

To eliminate the three fractions, first find a common denominator — that is, the least common multiple of 2, 4, and 6, which is 12. Multiply each term of the equation by this number:

$$12\left(\frac{3x}{4}\right)+12\left(\frac{7}{6}\right)=12\left(\frac{5x}{2}\right)$$

Now simplify by dividing 12 by the denominator and multiplying the result by the numerator. Do this in two steps so you don't get confused and make a mistake:

$$3(3x)+2(7)=6(5x)$$
$$9x+14=30x$$

At this point, you can solve the equation easily:

$$14=21x$$
$$\frac{14}{21}=x$$
$$\frac{2}{3}=x$$

Therefore, the correct answer is Choice (H).

Rooting out ways to solve square root equations

To solve an equation with a square root (or radical), isolate the root on one side of the equation. Then you can square both sides of the equation to remove the root.

REMEMBER

You must square the *entire* expression on each side of the equation. If you square each term individually, you'll get the wrong answer.

EXAMPLE

Which of the following values satisfies the equation $\sqrt{6x+17}-3=x$?

(A) $\sqrt{2}$

(B) $\sqrt{3}$

(C) $2\sqrt{2}$

(D) $2\sqrt{3}$

(E) $3\sqrt{2}$

To isolate the square root expression, add 3 to both sides of the equation:

$$\sqrt{6x+17}-3=x$$
$$\sqrt{6x+17}=x+3$$

To undo the square root, square both sides, using parentheses to make sure you square the *entire* expression on each side:

$$\left(\sqrt{6x+17}\right)^3=(x+3)^2$$

Now simplify. On the left side of the equation, the square undoes the square root, leaving you with the expression inside. On the right side, the square means you have to multiply the expression by itself and then FOIL it:

$$6x+17=(x+3)(x+3)$$
$$6x+17=x^2+6x+9$$

Simplify further, like so:

$$17=x^2+9$$
$$8=x^2$$

Take the square root of both sides and simplify again. Disregard the negative value $-\sqrt{8}$, because all five answers are positive:

$$\sqrt{8}=x$$
$$2\sqrt{2}=x$$

Therefore, the correct answer is Choice (C).

Gaining absolute confidence with absolute value

The *absolute value* of a number is its positive value, as I discuss in Chapter 4. Solving algebraic equations with absolute value presents a unique challenge. To unpack the value inside the absolute value bars, split the equation into two separate equations:

- The same equation with the bars removed
- The same equation with the bars removed *and* the opposite side of the equation negated

Splitting the equation usually results in two solutions rather than just one. In some cases, the question asks you for just one solution. In others, it may ask you to use both solutions to calculate the correct answer, as in the next example.

EXAMPLE

If $|3x+7|=5x-4$, what is the resulting product if you multiply all values of x that satisfy the equation?

(A) $-\frac{3}{22}$

(B) $\frac{8}{33}$

(C) $\frac{16}{33}$

(D) $-\frac{33}{16}$

(E) $\frac{44}{3}$

To begin, split the equation $|3x+7|=5x-4$ into two separate equations:

$$3x+7=5x-4 \qquad 3x+7=-(5x$$

Solve the first equation:

$$\begin{aligned}3x+7&=-(5x-4)\\7&=2x-4\\11&=2x\\\frac{11}{2}&=x\end{aligned}$$

Now solve the second equation:

$$\begin{aligned}3x+7&=-(5x-4)\\3x+7&=-5x-4\\8x+7&=4\\8x&=-3\\x&=-\frac{3}{8}\end{aligned}$$

Multiply these two solutions:

$$\left(\frac{11}{2}\right)\left(-\frac{3}{8}\right)=-\frac{33}{16}$$

So the correct answer is Choice (D).

Before splitting an equation, sometimes you may need to do some preliminary work to isolate the absolute value expression on one side of the equation. The following example shows you how.

EXAMPLE

Which of the following is a solution for the equation $\frac{|8k+2|}{k-3}=3$?

(F) 1

(G) $\frac{5}{11}$

(H) $\frac{7}{11}$

(J) $\frac{11}{5}$

(K) $\frac{11}{7}$

Your first step is to isolate $|8k+2|$ on one side of the equal sign. To do this, multiply both sides of the equation by $k-3$ and then simplify:

$$\begin{aligned}\frac{|8k+2|}{k-3}&=3\\|8k+2|&=3(k-3)\\|8k+2|&=3k-9\end{aligned}$$

Now you can split the equation:

$$8k+2=3k-9 \qquad 8k+2=-(3k-9)$$

Solve the first equation for k:

$$\begin{aligned}8k+2&=3k-9\\5k&=-11\\k&=-\frac{11}{5}\end{aligned}$$

This isn't one of the solutions given, so solve the second equation:

$$8k+2=-(3k-9)$$
$$8k+2=-3k+9$$
$$11k=7$$
$$k=\frac{7}{11}$$

Thus, the correct answer is Choice (H).

Exposing variables in the exponent

A tricky type of ACT question asks you to solve an equation with one or more variables in the exponent. For example, a question may ask you to solve for n in the following equation:

$$\left(\frac{1}{25}\right)^{n-1}=\sqrt{5}$$

At first glance, this problem appears unsolvable. But with a few tricks, you'll find that it's not so bad.

The main thing to know when solving this type of problem is how to express a variety of values as exponents with a common base. You can express whole numbers, fractions, and even square roots as exponents. For example, using a base of 5, Table 5-1 shows the numbers that you can express as exponents.

TABLE 5-1 Some Common Exponential Equivalents with a Base of 5

5^{-3}	5^{-2}	5^{-1}	5^{0}	5^{1}	5^{2}	5^{3}
$\frac{1}{125}$	$\frac{1}{25}$	$\frac{1}{5}$	1	5	25	125

And here's one more expression that you may need:

$$\sqrt{5}=5^{\frac{1}{2}}$$

EXAMPLE

If $\left(\frac{1}{25}\right)^{n-1}=\sqrt{5}$ then $n=$

(A) $\frac{1}{4}$

(B) $-\frac{1}{4}$

(C) $\frac{3}{4}$

(D) $-\frac{3}{4}$

(E) $\frac{5}{4}$

Your first goal is to express both sides of the equation in terms of a common base — in this case, 5. Replace both $\frac{1}{25}$ and $\sqrt{5}$ with their equivalents:

$$\left(\frac{1}{25}\right)^{n-1} = \sqrt{5}$$

$$\left(5^{-2}\right)^{n-1} = 5^{\frac{1}{2}}$$

On the right side of the equation, multiply the two exponents together to make a single exponent:

$$5^{-2(n-1)} = 5^{\frac{1}{2}}$$

This problem still doesn't look very easy to solve, but in fact you've made a great improvement: Because the bases on both sides of the equation are equal, the exponents also are equal. Therefore, you can throw away the bases and keep the exponents:

$$-2(n-1) = \frac{1}{2}$$

This equation is now pretty easy to solve for n. Begin by multiplying both sides by 2 to eliminate the fraction, and then distribute and solve:

$$-4(n-1) = 1$$
$$-4n + 4 = 1$$
$$-4n = -3$$
$$n = \frac{3}{4}$$

Therefore, the correct answer is (C).

EXAMPLE

What is the value of k if $\left(\frac{1}{8}\right)^{k} = 32$?

(F) $-\frac{2}{3}$

(G) $\frac{4}{3}$

(H) $-\frac{4}{3}$

(J) $\frac{5}{3}$

(K) $-\frac{5}{3}$

In this case, the numbers $\frac{1}{8}$ and 32 can both be expressed as powers of 2:

$$\frac{1}{8} = 2^{-3} \qquad 32 = 2^{5}$$

Substitute these two values into the equation:

$$\left(2^{-3}\right)^{k} = 2^{5}$$

Multiply the exponents on the right side of the equation:

$$2^{-3k} = 2^{5}$$

The two bases are equal, so the exponents also are equal; therefore, drop the bases and keep the exponents:

$$-3k = 5$$

Now this equation is easy to solve:

$$k = -\frac{5}{3}$$

The correct answer is Choice (K).

Taking extra care with extra variables

Generally speaking, when an equation has more than one variable, you can't solve it. However, some equations on the ACT may include one or more extra variables. In most cases, when you have an equation with extra variables, you must solve the equation *in terms of* the other variables — that is, isolate one variable on one side of the equation.

EXAMPLE

In the equation $2pq + 5qr = 3pr$, what is the value of p in terms of q and r?

(A) $\frac{2q+3r}{5qr}$

(B) $\frac{3r-2q}{5qr}$

(C) $\frac{5qr}{2q+3r}$

(D) $\frac{5qr}{2q-3r}$

(E) $\frac{5qr}{3r-2q}$

To answer this question, isolate p on one side of the equation. Begin by moving all the p terms to one side:

$$2pq + 5qr = 3pr$$
$$5qr = 3pr - 2pq$$

Now factor out p on the right side of the equation:

$$5qr = p(3r - 2q)$$

Divide both sides by $3r - 2q$:

$$\frac{5qr}{3r-2q} = p$$

So the correct answer is Choice (E).

Sometimes when an equation has more than one variable, one variable drops out of the equation (surprise!), allowing you to solve for the remaining variable. Check out the following example to see how.

EXAMPLE

What is the value of x in the equation $3(x+4y) = 6(2y+5)$?

(F) 10

(G) 14

(H) 17

(J) 23

(K) Cannot be determined from the information given

At first glance, the equation doesn't look solvable. But don't be too quick to jump to this conclusion. You can, in fact, solve this problem. Begin by distributing to remove the parentheses:

$$3(x+4y)=6(2y+5)$$
$$3x+12y=12y+30$$

Now subtract 12y from both sides of the equation:

$$3x=30$$

Magically, the y term has dropped out, leaving you with an equation that you can easily solve:

$$x=10$$

The correct answer is Choice (F).

Finally, you can sometimes solve an equation with extra variables for an expression that includes both variables.

EXAMPLE

If $\frac{2x-4}{y-1}=2$, what is the value of $x-y$?

(A) −2

(B) −1

(C) 1

(D) 2

(E) 4

This equation can't be solved for either x or y. However, you can isolate the expression $x-y$ on one side of the equation to answer the question. Begin by multiplying both sides by $y-1$:

$$\frac{2x-4}{y-1}=2$$
$$2x-4=2(y-1)$$

Now distribute on the right side of the equation:

$$2x-4=2y-2$$

Isolate the x and y terms on one side of the equation:

$$2x-2y-4=-2$$
$$2x-2y=2$$

Factor out a 2 from each term on the left side:

$$2(x-y)=2$$

Divide both sides by 2:

$$x-y=1$$

So the correct answer is Choice (C).

Lost in Translation: Translating Words into Equations

Some ACT questions give you information about what happens when you perform various operations on an unknown number. To answer these questions, let x (or any variable you choose) equal the number you're looking for. Then build an equation and solve it using any of the methods you already know.

EXAMPLE

When you divide a certain number by 2 and then add 2, the number you end up with is the same as when you add 24 and then divide this result by 3. What is the number?

(A) 30

(B) 36

(C) 42

(D) 48

(E) 56

To begin, let x be the number you're looking for. The first part tells you to divide x by 2 and then add 2:

$$\frac{x}{2}+2$$

The next part tells you to add 24 to x and then divide the result by 3:

$$\frac{x+24}{3}$$

The number you end up with is the same in both cases, so you can link these two expressions with an equal sign:

$$\frac{x}{2}+2=\frac{x+24}{3}$$

Now you have an equation that you can solve using any of the tools and tricks you know. Begin by multiplying every term here by the common denominator of 6 and simplify as I show you in the earlier section "Approaching rational equations rationally":

$$\begin{aligned}6\left(\frac{x}{2}\right)+6(2)&=6\left(\frac{x+24}{3}\right)\\3x+12&=2(x+24)\end{aligned}$$

Now simplify and solve for x:

$$\begin{aligned}3x+12&=2x+48\\x+12&=48\\x&=36\end{aligned}$$

So the correct answer is Choice (B).

IN THIS CHAPTER

» Understanding data presented in tables and graphs

» Working with Venn diagrams

» Finding the mean, median, and mode of a data set

» Calculating weighted averages

» Using the formula for probability

» Answering probability questions based on data presented in two-way frequency tables

Chapter 6
Real-World Math: Statistics and Probability

In this chapter, your study turns to statistics and probability, two ways that math can help you make sense of real-world observations presented in the form of data.

To begin, I give you practice reading data collected from tables and graphs. You also discover how to organize data in Venn diagrams. Next, I show you how to calculate the mean, median, and mode of data sets, and I walk you through the process of calculating weighted averages. From there, you work with the formula for probability and then apply it to data presented in two-way frequency tables.

Presenting Data in Tables and Graphs

Data is numerical information that's observed in the real world. This data is often presented in tables and graphs, so in this section I show you how to work with these visual representations. I also show you how to answer a variety of ACT questions that involve Venn diagrams.

Organizing information in tables

A *table* is a two-dimensional grid containing data that's organized into horizontal rows and vertical columns. A typical ACT question that includes a table may ask you to compare or calculate data, or it may ask you to fill in a missing value in the table. Often, one table will be used for more than one question.

Use the following table, which provides the gross sales amounts among six salespeople for four months (January through April), to answer the following two questions.

Team Leader	Salesperson	January	February	March	April
	Adam	89	41	71	65
Naomi	Patrick	49	59	70	55
	Stephanie	64	73	66	62
	Beth	71	72	63	65
Ronaldo	Carly	77	62	61	80
	Jefferson	81	61	66	56

Note: Figures in the table indicate thousands of dollars.

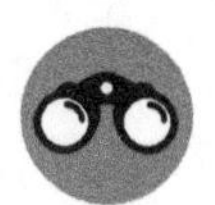

EXAMPLE

What is the total gross sales for Ronaldo's team in February?

(A) $171,000

(B) $190,000

(C) $195,000

(D) $201,000

(E) $239,000

Ronaldo's team includes Beth, Carly, and Jefferson, whose February sales totaled $72,000 + $62,000 + $61,000 = $195,000. Therefore, the correct answer is Choice (C).

EXAMPLE

The first quarter is January through March, inclusive. If the quarterly sales quota is $180,000, which of the following salespeople failed to make their first-quarter quota?

(F) Adam

(G) Beth

(H) Carly

(J) Jefferson

(K) Patrick

You can add up the numbers for the first three months for each of the five salespeople, but there's a quicker way. Notice that the quota of $180,000 over three months requires a monthly average of $60,000. A quick look at the table shows that Patrick made less than $60,000 for two of the three months, so start with him. Patrick's sales for January through March were $49,000 + $59,000 + $70,000 = $178,000, so the correct answer is Choice (K).

Representing data with graphs

A *graph* is a visual representation of a data set. Some common types of graphs include the following:

- **Bar graphs,** which show comparisons among numerical data
- **Pie charts,** which show data represented as percentages of a whole
- **Line graphs,** which show changes that occur over time
- **Scatterplots,** which show general data trends by providing a large variety of data in two dimensions
- **Pictograms,** which show comparisons among numerical data in discrete increments

Graphs tend to be self-explanatory — that's why they're so useful. If you're clear on what the numbers in a graph mean, you'll probably be able to answer just about any question that the ACT throws at you.

In this section, I provide questions that include a variety of graphs. Additionally, the questions in Chapter 7 give you more practice working with graphs.

The following pictogram shows the population of five neighboring counties. Which county has a population of 10,000?

Abercrombie 🯅
Brandt 🯅 🯅
Closter 🯅 🯅 🯅
Dedham 🯅 🯅 🯅 🯅 🯅
Elkinsboro 🯅 🯅 🯅 🯅 🯅 🯅 🯅 🯅 🯅 🯅

🯅 = 5,000 people

(A) Abercrombie

(B) Brandt

(C) Closter

(D) Dedham

(E) Elkinsboro

Each figure represents 5,000 people, so two figures represents 10,000. As you can see, Brandt county has 10,000 people, so the correct answer is Choice (B).

EXAMPLE

The following bar graph shows the number of pots of coffee brewed by the Filterfresh Coffee-house on each day last week. On which day did they brew exactly 17 pots of coffee?

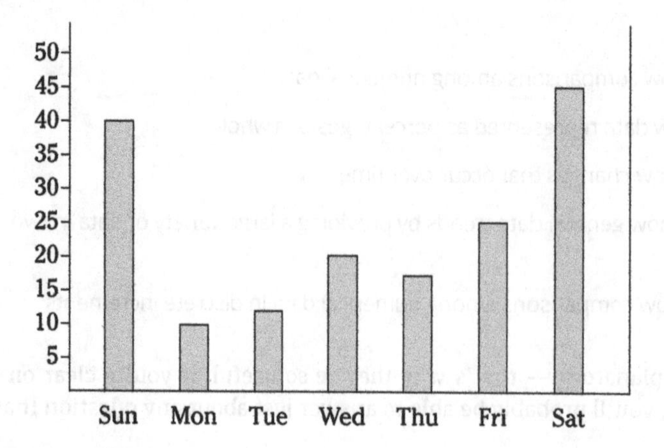

(F) Monday
(G) Tuesday
(H) Wednesday
(J) Thursday
(K) Saturday

The coffeehouse brewed exactly 17 pots of coffee on Thursday, so the correct answer is Choice (J).

EXAMPLE

The following pie chart shows how \$500,000 of funding for an organization will be allocated. Which of the following pairs of services will account for exactly \$275,000 of the funding allocated?

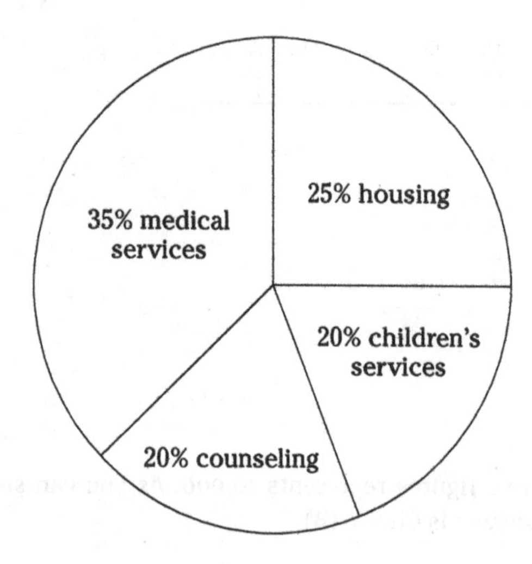

(A) Children's services and housing
(B) Counseling and children's services
(C) Counseling and housing
(D) Counseling and medical services
(E) Housing and medical services

Because the pie chart divides the $500,000 by percentages, you need to figure out how much each slice of the pie is worth in dollars. Children's services and counseling are both 20%, so each of these slices represents $100,000. Housing is 25%, so this slice represents $125,000. Medical services represents the remaining funding, so it represents $175,000. The combination of counseling and medical services represents $\$100,000 + \$175,000 = \$275,000$, so the correct answer is Choice (D).

Understanding Venn diagrams

A *Venn diagram* is a visual tool that uses circles for organizing data. The simplest Venn diagram has a pair of circles, each of which represents a different subgroup within a larger group. For example:

EXAMPLE

The graduating class of Rutherford B. Hayes High School has a total of 78 students. Of these, 18 are on the honor roll and 32 are on at least one sports team. If 37 graduating students aren't on the honor roll and don't play a sport, how many honor-roll students play at least one sport?

(A) 4

(B) 9

(C) 14

(D) 19

(E) 24

You can use a Venn diagram to help you organize this information:

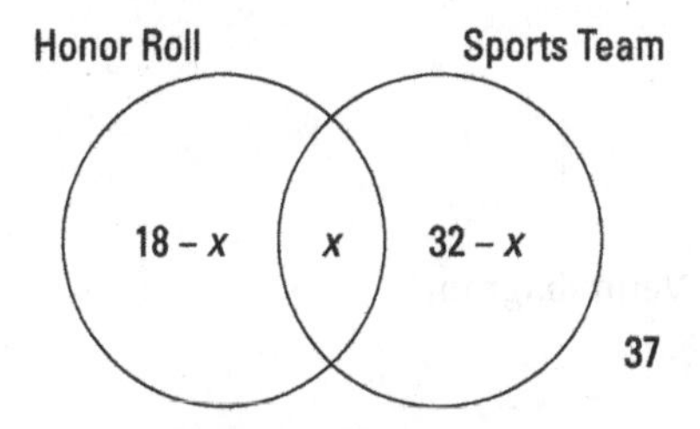

In this diagram, the circle on the left contains the total number of honor-roll students, and the circle on the right contains the total number of students who are on at least one sports team. The overlapping area between these two circles tells you the number of students who play on both teams, and the area outside the circles but inside the rectangle indicates the 37 students who belong to neither group.

As you can see, I've filled in x inside the overlapping area. Then, because a total of 18 graduating students are on the honor roll, I've filled in $18 - x$ into the area of the left circle. This indicates students who are *only* on the honor roll. I filled in $32 - x$ into the area of the right circle, which represents students who are *only* on a sports team. The total of these four groups equals 78, so I make and solve the following equation:

$$\begin{aligned} 18 - x + x + 32 - x + 37 &= 78 \\ 87 - x &= 78 \\ -x &= -9 \\ x &= 9 \end{aligned}$$

Therefore, 9 graduating students are both on the honor roll and on at least one sports team. The correct answer is Choice (B).

You can also use a Venn diagram with three circles to organize complex information that would be very difficult to understand without such a chart. For example:

At Chester A. Arthur High School, the three most popular extracurricular activities are the Drama Club, the Lacrosse Team, and the Marching Band. Among the students at the high school:

45 belong to the Drama Club

31 belong to the Lacrosse Team

38 belong to the Marching Band

19 belong to both the Drama Club and the Lacrosse Team

22 belong to both the Drama Club and the Marching Band

15 belong to both the Lacrosse Team and the Marching Band

11 belong to all three of these clubs

29 belong to none of these clubs

How many students attend Chester A. Arthur High School?

(F) 98
(G) 114
(H) 137
(J) 165
(K) 210

To organize this information, make the following Venn diagram:

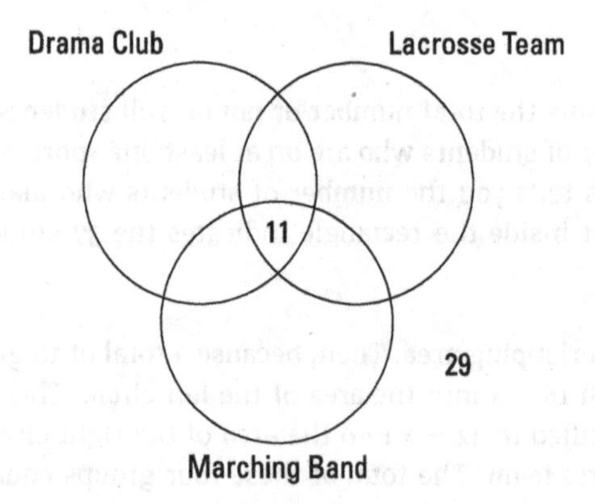

Notice that to start, I've filled two numbers into the chart: the 11 students who are in all three clubs and the 29 students who aren't in any clubs. Next, I fill in the students who are in exactly two clubs as follows:

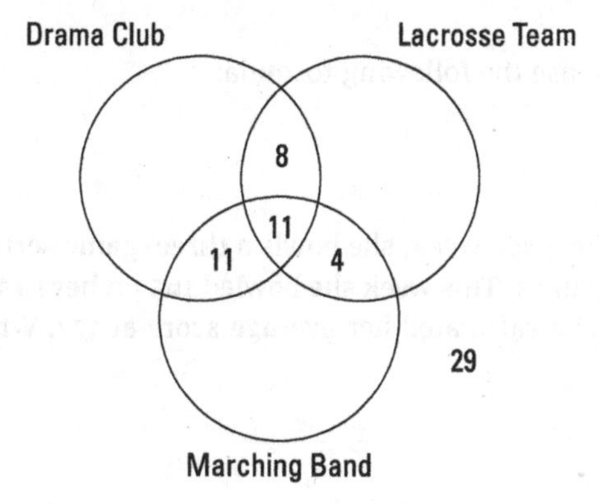

As you can see, I've filled in numbers that correctly represent the three groups of students who are in two or more of the clubs. To finish up, I complete the chart, using the numbers of students who are in at least one club:

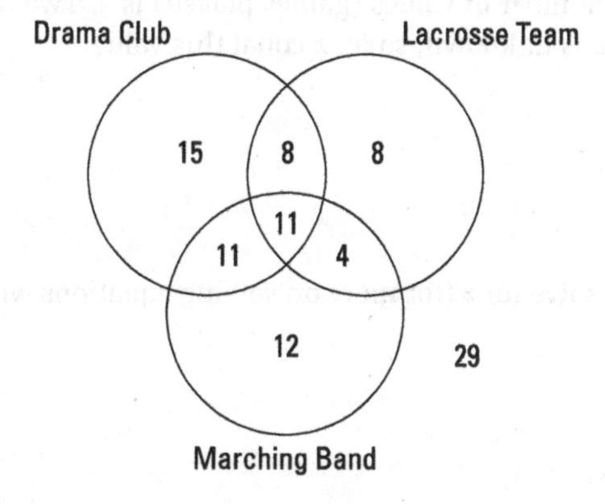

To complete the problem, add up all the values in the Venn diagram:

$$15+8+12+8+11+4+11+29=98$$

Therefore, the correct answer is Choice (F).

Getting a Better-Than-Average Grasp of Averages

ACT questions involving statistics often require you to calculate one of three types of averages for a data set: the mean, the median, and the mode. In this section, I show you how to calculate all three types of averages. I also give you practice calculating weighted averages.

Finding the meaning of the mean

The most common type of average is the *mean* (also called the *arithmetic mean* and the *mean average*). When an ACT question asks for an average and doesn't specify which type, assume that the question is asking for the mean.

To calculate the mean of a list of numbers, use the following formula:

$$\text{Mean} = \frac{\text{Sum of values}}{\text{Number of values}}$$

Sasha bowls every week in a bowling league. Each week, she bowls a three-game series and then calculates the average score for her three games. This week she bowled 165 on her first game and 175 on her second. After the third game, she calculated her average score at 172. What was her score on the third game?

(A) 172

(B) 173

(C) 174

(D) 175

(E) 176

In this question, the mean is 172 and the number of values (games played) is 3. Two of the three values are 165 and 170, and the third value is unknown, so let *x* equal this value:

$$\text{Mean} = \frac{\text{Sum of values}}{\text{Number of values}}$$
$$172 = \frac{165 + 175 + x}{3}$$

To find the remaining value, simplify and solve for *x* (for more on solving equations with algebra, see Chapter 5):

$$172 = \frac{340 + x}{3}$$
$$516 = 340 + x$$
$$176 = x$$

Sasha bowled 176 on her third game, so the correct answer is Choice (E).

In some cases, you may need to find the average of a set of values that includes one or more variables. The following example shows you how.

If the average (arithmetic mean) of *p*, 9*p*, 10*p*, and 19 is 6, what is the value of *p*?

(F) $\frac{1}{2}$

(G) $\frac{1}{3}$

(H) $\frac{2}{3}$

(J) $\frac{1}{4}$

(K) $\frac{3}{4}$

In this case, the mean is 6 and you're given all four values, so you simply plug these into the formula:

$$\text{Mean} = \frac{\text{Sum of values}}{\text{Number of values}}$$
$$6 = \frac{p + 9p + 10p + 19}{4}$$

Simplify and solve for p (refer to Chapter 5 as necessary to get help with the algebra):

$$6 = \frac{20p + 19}{4}$$
$$24 = 20p + 19$$
$$5 = 20p$$
$$\frac{1}{4} = p$$

As you can see, the correct answer is Choice (J).

Centering on the median

Another important type of average is the *median*, which is the middle number in a list of numbers that are in sequential order. When a list contains an even number of values, the median is the arithmetic mean of the two middle numbers.

EXAMPLE

The Garfield Alternative School has six classrooms for students from kindergarten through fifth grade. The current enrollment for these six classes is shown in the following table. What is the median number of students in the six classes?

Kindergarten	*1st Grade*	*2nd Grade*	*3rd Grade*	*4th Grade*	*5th Grade*
26	24	30	31	23	33

(A) 25

(B) 26

(C) 27

(D) 28

(E) 29

Place the six values in ascending order: 23, 24, 26, 30, 31, 33. The two middle values are 26 and 30, so you have to find the mean of these two numbers:

$$\text{Mean} = \frac{\text{Sum of values}}{\text{Number of values}} = \frac{26 + 30}{2} = \frac{56}{2} = 28$$

The median is 28, so the correct answer is Choice (D).

Discovering what's so popular about the mode

A third type of average is the *mode*, which is the most frequently-appearing number in a list of numbers. To find the mode, you simply have to count which number occurs most in a list. Pretty easy, huh?

EXAMPLE

Geoff played golf every day last week. His scores were 85 on Sunday, 83 on Monday, 85 on Tuesday, 82 on Wednesday, 86 on Thursday, 83 on Friday, and 83 on Saturday. What were his median and mode scores for the week?

(A) Median = 83, Mode = 83

(B) Median = 83, Mode = 84

(C) Median = 83, Mode = 85

(D) Median = 84, Mode = 83

(E) Median = 84, Mode = 85

To find the median, put the seven scores in order from lowest to highest, like so:

82, 83, 83, 83, 85, 85, 86

The median score, as I explain in the previous section "Centering on the median," is the number that appears in the middle of the list. In this problem, it's 83. The mode score is the one that appears most frequently in the list, so it's also 83. Thus, the correct answer is Choice (A).

Weighing in on weighted averages

A *weighted average* (or *weighted arithmetic mean*) is an average that takes into account the relative importance of values in a data set. For example:

EXAMPLE

Carrie-Ann's chemistry grade is calculated as 30% of the average of her lab scores plus 70% of the average of her test scores. Her five scores as of October 15 are shown in the following table. On this basis, what is her current chemistry grade, rounded to the nearest whole point?

Lab #1	95
Lab #2	92
Lab #3	85
Test #1	89
Test #2	97

(A) 88

(B) 89

(C) 90

(D) 91

(E) 92

To calculate Carrie-Ann's grade, first find the averages of her lab and test scores:

$$\text{Lab Average} = \frac{95+92+85}{3} = \frac{272}{3} \approx 90.7 \qquad \text{Test Average} = \frac{89+97}{2} = \frac{186}{3} \approx 93$$

Now, calculate her grade as follows:

$$\text{Grade} = .3(90.7) + .7(93) \approx 27.2 + 65.1 = 92.3$$

Therefore, her grade is approximately 92, so the correct answer is Choice (E).

Looking at Likelihood: Probability

Probability is the mathematical likelihood that something will happen. ACT questions about probability usually involve certain common scenarios such as tossing a coin, picking a card, or choosing an item at random. I help you become familiar with probability in the following sections.

Learning to count

Calculating probability requires a systematic way to count *outcomes*, which are the possible ways in which a set of events can take place. For example, if you toss four coins, one possible outcome is that all four come up heads. All four of these events are *independent* because what happens to one coin doesn't affect any of the others.

But if you pick three letters from a bag that contains one copy of each letter from A to Z, one possible outcome is picking the letters A, M, and Z. In this case, however, the three events are *dependent*, because if you pick A first, you can't also pick it second or third.

In the following sections, I show you how to count outcomes of both *independent events* (events that don't affect each other) and *dependent events* (events that affect each other).

Counting the outcomes of independent events

When two (or more) events are *independent*, the outcome of one event has no effect on the outcome of the other. Tossing two coins and rolling a pair of dice are considered independent events, because what happens to one coin or die doesn't affect what happens to the other. Counting independent events is an important first step to calculating probability, and it can also help you directly when answering some ACT questions.

EXAMPLE

If you toss ten coins into the air, in how many distinct ways can they land?

(A) Fewer than 100

(B) Between 100 and 200

(C) Between 200 and 500

(D) Between 500 and 1,000

(E) More than 1,000

In this question, how each coin lands (either heads or tails) is an *event*. These events don't affect each other — that is, how one coin lands has no effect on any of the other coins — so they're *independent events*. To count the number of possible outcomes for these ten events, track each coin separately as follows:

#1	*#2*	*#3*	*#4*	*#5*	*#6*	*#7*	*#8*	*#9*	*#10*
2	2	2	2	2	2	2	2	2	2

Each coin, #1 through #10, can land in one of two ways: either heads or tails. To calculate the number of possible outcomes, multiply 2 by itself 10 times. A fast way to do this on your calculator is 2^{10}:

$$2^{10} = 1{,}024$$

The correct answer is Choice (E).

Counting the outcomes of dependent events

When two (or more) events are *dependent*, the outcome of one event affects the outcome of the other. Picking letters from a bag and socks from a drawer are dependent events: The outcome of the first event affects the outcome of the second event, because after you pick a letter or sock, you can't pick the same one again. Identifying dependent events and counting them correctly is important for calculating probability.

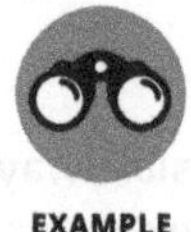

EXAMPLE

Alex Ward Patton noticed that his own initials (AWP) contain no repeated letters, but the initials of his best friend, James Dean Jackson (JDJ), contain repeated letters. How many different sets of three initials have no repeating letters?

(A) 650

(B) 676

(C) 15,600

(D) 15,625

(E) 17,576

In this question, each initial is a different event. Because no initial can be repeated, each event affects the other events. For example, if a first initial is A, then the second initial can't also be A. Thus, the number of possible first initials is 26, the number of possible second initials is 25, and the number of possible third initials is 24.

To find the number of possible outcomes, you simply multiply these three numbers together:

$$26 \times 25 \times 24 = 15{,}600$$

You can see that the correct answer is Choice (C).

Determining probability

Probability, the mathematical likelihood of an outcome, is always a number from 0 to 1. When the probability of an outcome is 0, the outcome is impossible; when the probability is 1, the outcome is certain.

Here's the formula for probability:

$$\text{Probability} = \frac{\text{Target outcomes}}{\text{Total outcomes}}$$

In this formula, you replace *target outcomes* with the number of ways in which a specific outcome occurs. You replace *total outcomes* with the number of ways in which *any* outcome can occur.

EXAMPLE

If you toss two coins, what is the probability that at least one of them will land heads up?

(A) $\frac{1}{2}$

(B) $\frac{1}{3}$

(C) $\frac{2}{3}$

(D) $\frac{1}{4}$

(E) $\frac{3}{4}$

To find the total number of outcomes, count the number of events. When doing so, you discover that the first coin has two and the second coin has two. Now multiply these numbers together (pretty tough, isn't it?):

$$2 \times 2 = 4$$

So you can assume 4 total outcomes, as follows:

HH HT TH TT

Of these, 3 outcomes have at least one heads, so the target outcomes is 3. Plug these numbers into the formula for probability:

$$\text{Probability} = \frac{\text{Target outcomes}}{\text{Total outcomes}} = \frac{3}{4}$$

The probability that at least one coin lands heads up is three-fourths, so the correct answer is Choice (E).

EXAMPLE

If a bag contains four white socks and four black socks, what is the probability of pulling three black socks at random from the bag?

(F) $\frac{1}{8}$

(G) $\frac{3}{8}$

(H) $\frac{1}{14}$

(J) $\frac{3}{14}$

(K) $\frac{1}{24}$

To calculate the probability, you need to calculate the total outcomes and the target outcomes. Keep in mind that the question involves dependent events, because if you pull a certain sock out of the bag first, you can't pull it out of the bag second or third.

First, count the total outcomes: You can pull eight possible socks out first, any of the remaining seven socks second, and any of the remaining six socks third. Multiply to find the total outcomes:

$$8 \times 7 \times 6 = 336$$

So this scenario has a total of 336 possible outcomes.

Next, count up the target outcomes: You can pull any of the four black socks out first, any of the remaining three black socks second, and either of the remaining two black socks third. Multiply to find the target outcomes:

$$4 \times 3 \times 2 = 24$$

Now, to get your answer, simply plug the numbers into the formula for probability:

$$\text{Probability} = \frac{\text{Target outcomes}}{\text{Total outcomes}} = \frac{24}{336} = \frac{1}{14}$$

The correct answer is Choice (H).

Solving more complex probability problems

More complex ACT questions that involve probability may require you to think critically about how words such as *not*, *or*, *and*, and *if* affect a probability calculation. In this section, I give you a few key tools for unraveling such problems.

Knowing how to take a complement

When calculating probability on the ACT, you may need to calculate the probability of the *complement* of an event — that is, the probability that the event will *not* occur. To calculate this value:

1. **Find the probability that the event *will* occur.**
2. **Subtract this value from 1.**

For example:

EXAMPLE

Shauna hopes to participate in a class trip. Unfortunately, the school bus can accommodate only 16 of the 28 students who want to go on the trip. To be fair, her teacher is holding a lottery to randomly pick the 16 students. What is the probability that Shauna will *not* be chosen to go on the trip?

(A) $\frac{2}{7}$

(B) $\frac{3}{7}$

(C) $\frac{4}{7}$

(D) $\frac{5}{7}$

(E) $\frac{6}{7}$

To begin, calculate the probability that Shauna will be randomly chosen to go on the trip as follows:

$$\text{Probability} = \frac{\text{Target outcomes}}{\text{Total outcomes}} = \frac{16}{28} = \frac{4}{7}$$

Now, to find the event that Shauna will not be chosen, subtract this value from 1:

$$1 - \frac{4}{7} = \frac{3}{7}$$

So the correct answer is Choice (B).

Calculating probabilities using the words "and" and "or"

Probability questions involving the words *and* and *or* can be tricky. You need to read them carefully in order to fully understand what they're asking, so you can calculate the correct probability. For example:

EXAMPLE

Stephen placed 20 tiles in a bag, each with a different integer from 1 to 20. He pulled one tile from the bag at random. What is the probability that the tile he chose was *both* a prime number *and* a number greater than 10?

(A) $\frac{1}{4}$

(B) $\frac{1}{5}$

(C) $\frac{2}{5}$

(D) $\frac{3}{10}$

(E) $\frac{7}{10}$

To begin, list the prime numbers and the numbers greater than 10:

Prime numbers: 2, 3, 5, 7, 11, 13, 17, 19

Greater than 10: 11, 12, 13, 14, 15, 16, 17, 18, 19, 20

The question is asking you to identify the numbers that are in *both* sets:

Prime numbers that are greater than 10: 11, 13, 17, 19

Thus, 4 out of the 20 tiles fulfill the condition, so calculate as follows:

$$\text{Probability} = \frac{\text{Target outcomes}}{\text{Total outcomes}} = \frac{4}{20} = \frac{1}{5}$$

Therefore, the correct answer is Choice (B).

Here's a question that includes a subtle and important difference from the previous one:

EXAMPLE

Stephen placed 20 tiles in a bag, each with a different integer from 1 to 20. He pulled one tile from the bag at random. What is the probability that the tile he chose was *either* a prime number *or* a number greater than 10, or both?

(F) $\frac{1}{4}$

(G) $\frac{1}{5}$

(H) $\frac{2}{5}$

(J) $\frac{3}{10}$

(K) $\frac{7}{10}$

In this case, the question is asking you to identify the numbers that are in *either* of the two sets:

Prime numbers *or* greater than 10: 2, 3, 5, 7, 11, 12, 13, 14, 15, 16, 17, 18, 19, 20

Thus, 14 out of the 20 tiles fulfill the condition, so calculate as follows:

$$\text{Probability} = \frac{\text{Target outcomes}}{\text{Total outcomes}} = \frac{14}{20} = \frac{7}{10}$$

So, the correct answer is Choice (K).

Knowing about conditional probability

Conditional probability is the probability that an outcome will occur *given that* an initial condition is met. In most cases, conditional probability shrinks the pool of total outcomes. An example here should be helpful:

EXAMPLE

Sandra is playing a game that utilizes an eight-sided die with faces numbered 1 to 8. In order to move her first piece, she must roll an odd number. What are the odds that her first roll that allows her to move will be the number 7?

(A) $\frac{1}{8}$

(B) $\frac{1}{4}$

(C) $\frac{1}{2}$

(D) $\frac{3}{4}$

(E) $\frac{7}{8}$

Sandra can move only if she rolls an odd number — that is, 1, 3, 5, or 7. Thus, the pool of total outcomes is reduced from 8 possibilities to 4. Of these, only one (rolling a 7) is the target outcome. Therefore, calculate the probability as follows:

$$\text{Probability} = \frac{\text{Target outcomes}}{\text{Total outcomes}} = \frac{1}{4}$$

Therefore, the correct answer is Choice (B).

Working with contingency tables

Contingency tables provide information based on two different attributes within a data set.

In this section, use the information and the contingency table to answer the three questions that follow.

A building has 139 one-bedroom and two-bedroom condominiums, some of which have views of the nearby bay. This data is organized into the following contingency table:

	Bay View	No Bay View	Total
1 Bedroom	14	63	77
2 Bedrooms	27	35	62
Total	41	98	139

EXAMPLE

What is the probability that a condominium selected at random will be a one-bedroom unit *and* will have a bay view?

(A) $\frac{14}{41}$

(B) $\frac{27}{41}$

(C) $\frac{14}{139}$

(D) $\frac{27}{139}$

(E) $\frac{41}{139}$

According to the table, of the 139 condominiums, 14 are one-bedroom units that have a bay view, so calculate this probability as follows:

$$\text{Probability} = \frac{\text{Target outcomes}}{\text{Total outcomes}} = \frac{14}{139}$$

Therefore, the correct answer is Choice (C).

EXAMPLE

What is the probability that a condominium selected at random will either be a two-bedroom unit with a bay view *or* a one-bedroom unit without a bay view?

(F) $\frac{41}{139}$

(G) $\frac{77}{139}$

(H) $\frac{90}{139}$

(J) $\frac{104}{139}$

(K) $\frac{118}{139}$

According to the table, of the 139 condominiums, 27 are two-bedroom units that have a bay view, and 63 are one-bedroom units without a bay view, so calculate this probability as follows:

$$\text{Probability} = \frac{\text{Target outcomes}}{\text{Total outcomes}} = \frac{27+63}{139} = \frac{90}{139}$$

Therefore, the correct answer is Choice (H).

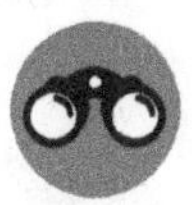

EXAMPLE

What is the probability that a one-bedroom condominium selected at random will have a bay view?

(A) $\frac{2}{5}$

(B) $\frac{2}{7}$

(C) $\frac{2}{11}$

(D) $\frac{5}{11}$

(E) $\frac{7}{11}$

According to the table, of the 77 one-bedroom condominiums, 14 have a bay view, so calculate this probability as follows:

$$\text{Probability} = \frac{\text{Target outcomes}}{\text{Total outcomes}} = \frac{14}{77} = \frac{2}{11}$$

Therefore, the correct answer is Choice (C).

Chapter 7

Practice Problems for Pre-Algebra and Elementary Algebra

Ready for some pre-algebra and elementary algebra practice? You've come to the right place! In this chapter, I provide 36 practice problems that include everything covered in Chapters 4, 5, and 6. If you get stuck, each answer is completely worked out in the later section titled (you guessed it) "Solutions to Practice Problems." Now get started!

Practice Problems

Following are 36 problems for you to practice your pre-algebra and elementary algebra skills. If you need to brush up on any of the concepts, refer to Chapters 4, 5, and 6.

1. If the first six numbers of a sequence are, in order, 6, 12, 9, 18, 15, 30, what is the eighth number in the sequence?

   (A) 36

   (B) 42

   (C) 45

   (D) 48

   (E) 54

2. Danielle spends her summers working as a baby sitter for families in the neighborhood. She began one day with $20, and then she worked for 4 hours in the morning at $10 per hour and received a $5 tip. In the afternoon, she worked for 3 more hours at $12 per hour but needed to pay $7 to take a cab home. How much money did she have at the end of the day?

(F) $74

(G) $94

(H) $104

(J) $108

(K) $118

3. A charitable organization sells books of raffle tickets as a fundraiser. The following graph shows the number of books that five different people have sold so far. What is the average number of books sold among these five people?

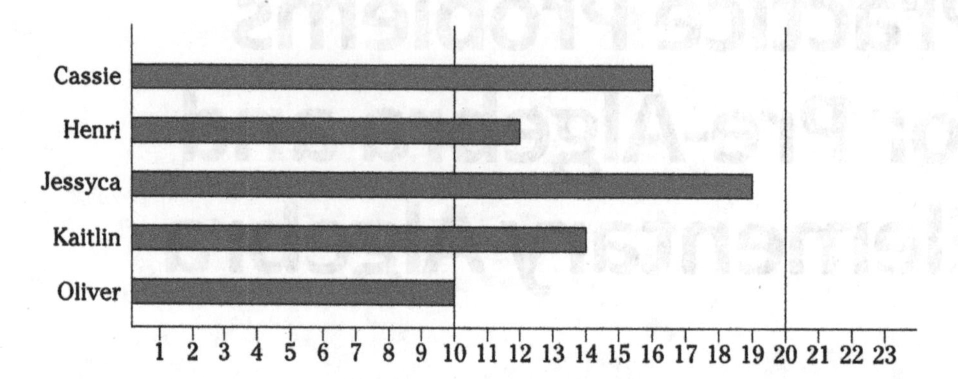

(A) 14

(B) 14.2

(C) 14.4

(D) 14.6

(E) 15

4. If $5(x-2)=3(2x-6)$, then $x=$

(F) 7

(G) 8

(H) 9

(J) 11

(K) 13

5. What is the value of $n^4+5kn-k^3$ if $k=-2$ and $n=3$?

(A) 51

(B) 53

(C) 55

(D) 57

(E) 59

6. If you subtract 1 from a real number and then multiply by 5, the result is the same as if you multiply the same real number by 4 and then add 0.5. What is the real number?

(F) 4.5

(G) 5.5

(H) 6.25

(J) 7.25

(K) 8

7. If $\frac{n+4}{9}=\frac{6-n}{11}$, then $13n-2n^2=$

(A) 6

(B) −8

(C) 10

(D) 12

(E) −14

8. If $ax+by=c$, then $x=$

(F) $aby+ac$

(G) $ac-aby$

(H) $\frac{by+c}{a}$

(J) $\frac{c-by}{a}$

(K) $\frac{b-c}{a}y$

9. What is the value of $\frac{10j+10k}{j^2-k^2}$ if $j-k=100$?

(A) 0.01

(B) 0.1

(C) 1

(D) 10

(E) 100

10. While packing for a business trip, Kwame found that he had room for 4 suits, 6 shirts, and 12 ties. Assuming that he can wear any combination of one suit, one shirt, and one tie, how many combinations are possible?

(F) 4

(G) 12

(H) 22

(J) 288

(K) 4,096

11. If $\frac{n}{10} - \frac{1-n}{2} = \frac{n+3}{5}$, which of the following is the value of n?

(A) $\frac{11}{2}$

(B) $\frac{11}{3}$

(C) $\frac{11}{4}$

(D) $\frac{11}{5}$

(E) $\frac{11}{6}$

12. The following graph shows the results of a poll asking people how many pets they have. What percentage of the people asked have three or more pets?

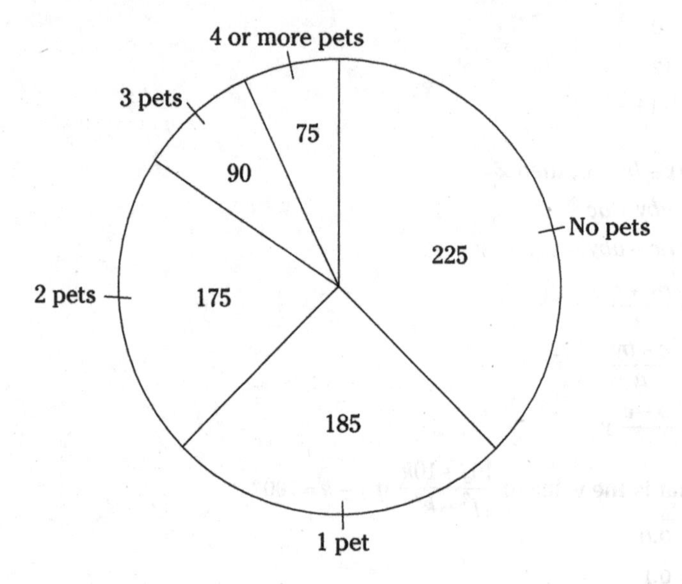

(F) 20 percent

(G) 21 percent

(H) 22 percent

(J) 22.5 percent

(K) 25 percent

13. If you double a number and then subtract 6, the result is the same as if you add 18 to the same number and then multiply what you get by –4. What is the result when you add 14 to the number?

(A) –13

(B) –3

(C) 3

(D) 13

(E) 23

14. If $9^{m-3} = \sqrt{3}$, then $m =$

(F) $\frac{7}{2}$

(G) $\frac{7}{4}$

(H) $-\frac{7}{4}$

(J) $-\frac{11}{4}$

(K) $\frac{13}{4}$

15. If $\frac{4x-3y}{2x-9} = 2$, what is the value of y?

(A) 6

(B) −6

(C) 15

(D) −15

(E) Cannot be determined from the information given

16. What is the positive value of p if $|17-2p| = 25$?

(F) 4

(G) 8

(H) 15

(J) 21

(K) 42

17. How many factors does the number 84 have?

(A) 6

(B) 8

(C) 9

(D) 10

(E) 12

18. What is the greatest common factor among $8x^3y^2z^4$, $12x^2y^3z^5$, and $20xy^6z^7$?

(F) $2y^2z^4$

(G) $2x^2y^2z^4$

(H) $4xy^2z^4$

(J) $4xy^3z^4$

(K) $4x^3y^6z^7$

19. Eleven children took a test and scored 78, 83, 83, 84, 88, 91, 93, 93, 93, 95, and 96. Which of the following statements is true?

(A) The median is two points higher than the mode.

(B) The median is one point higher than the mode.

(C) The median and the mode are the same.

(D) The median is one point lower than the mode.

(E) The median is two points lower than the mode.

20. Marv bought a flat-screen TV that was marked at 20 percent off its original price. The store added 10 percent tax to the marked-down price, so Marv ended up paying $880. What was the original price of the TV?

(F) $800

(G) $880

(H) $968

(J) $1,000

(K) $1,080

21. If m is the lowest common denominator of the fractions $\frac{5}{6}$, $\frac{3}{8}$, and $\frac{1}{9}$, which of the following is true?

(A) $m < 50$

(B) $50 \leq m < 60$

(C) $60 \leq m < 70$

(D) $70 \leq m < 80$

(E) $m \geq 80$

22. What product results when you multiply the two solutions of $|7n + 1| = 3n - 11$?

(F) 3

(G) –3

(H) –6

(J) 9

(K) –12

23. A school has a 2:3:35 ratio of administrators, teachers, and students, respectively. If the school has exactly 15 teachers, what is the total number of administrators, teachers, and students?

(A) 160

(B) 175

(C) 200

(D) 320

(E) 375

24. If you divide a number by 3 and then add 13, the number you end up with is the same as if you subtract 36 from the same number and then multiply by 2. What is the number?

(F) 51

(G) 63

(H) 75

(J) 81

(K) 99

The following graph shows the number of houses Lakshmi sold in each of the first six months of the year. Use this graph to answer Questions 25 and 26.

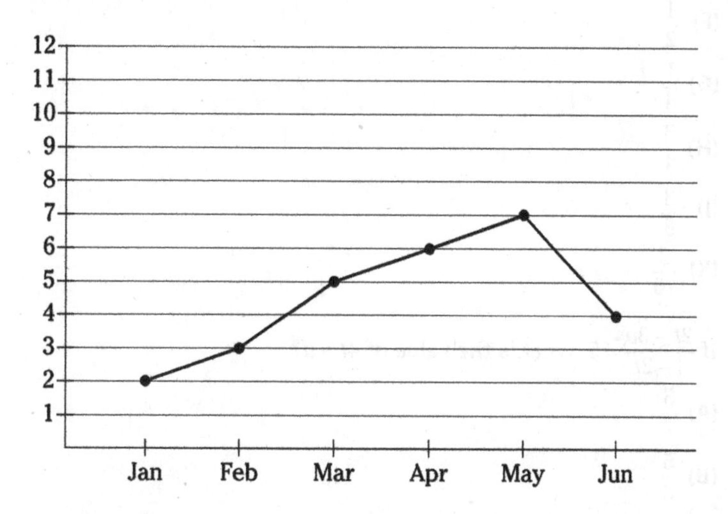

25. What is the total number of houses Lakshmi sold from January through March, inclusive?

(A) 5

(B) 7

(C) 8

(D) 10

(E) 12

26. What is the average number of houses that Lakshmi sold over the six months?

(F) 4

(G) 4.5

(H) 5

(J) 5.5

(K) 6

27. If $\sqrt{x^2+1}-x=2$, then $x=$

(A) $\frac{3}{4}$

(B) $-\frac{3}{4}$

(C) $\frac{5}{4}$

(D) $-\frac{5}{4}$

(E) $\frac{7}{4}$

28. If you toss a penny, a nickel, a dime, and a quarter, what is the probability that exactly two of the coins will land heads up and exactly two will land tails up?

(F) $\frac{1}{2}$

(G) $\frac{1}{4}$

(H) $\frac{3}{4}$

(J) $\frac{3}{8}$

(K) $\frac{3}{16}$

29. If $\frac{2t-3u}{1-2t}=5$, what is the value of $4t-u$?

(A) $\frac{5}{2}$

(B) $\frac{5}{3}$

(C) $\frac{10}{3}$

(D) $\frac{11}{2}$

(E) Cannot be determined from the information given

30. In his first four basketball games this season, Jerome has scored 13, 17, 19, and 21 points. How many points does he need to score in his next game to bring his average score for the first five games up to 20 points per game?

(F) 10

(G) 20

(H) 26

(J) 27

(K) 30

31. The art department at Michelangelo University currently has 75 students, 15 of whom are both painters and sculptors. The number of students who are painters is twice the number of students who are sculptors. If 12 of the art students are neither painters nor sculptors, how many of these students are painters but not sculptors?

(A) Fewer than 12

(B) From 12 to 20

(C) From 21 to 30

(D) From 31 to 40

(E) More than 40

Each of the 151 employees of J. Stone Construction, LLC, either management or union members, works either at the Branch Avenue or the Wallace Street location. The following contingency table provides a detailed accounting of this data:

	Management	Union Member	Total
Branch Avenue	18	47	65
Wallace Street	25	61	86
Total	43	108	151

Use this information to answer Questions 32 and 33.

32. **Approximately what percentage of the people who work at the Wallace Street location are union members?**

(F) 41%

(G) 57%

(H) 58%

(J) 71%

(K) 72%

33. **The circle graph below is a visual representation of the data found in the table above. In this circle graph, the two sections labeled ~17% and ~31% together account for which of the following subgroups at the company?**

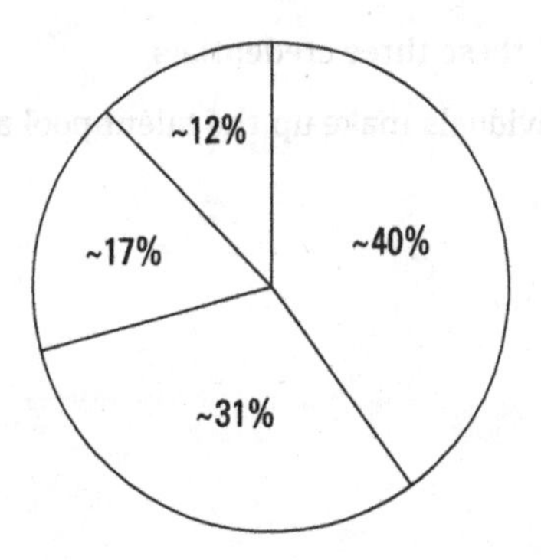

(A) All management at both locations

(B) All union members at both locations

(C) All employees at the Wallace Street location

(D) Branch Avenue management and Wallace Street union members

(E) Branch Avenue union members and Wallace Street management

34. **Of the 24 students in Mr. Johnson's math class last semester, the 7 who received As for their final grade had an average score of 94 on the final exam, the 12 who got Bs had an average score of 85, and the 5 who got Cs had an average score of 77. Which of the following was closest to the average score on the final exam for the whole class?**

(F) 82

(G) 83

(H) 84

(J) 85

(K) 86

35. Erin and Amara are 2 of the 7 candidates who are currently applying for a grant to visit Japan this summer to do research. If only 2 applicants of the 7 will be selected, what is the probability that the two applicants will be Erin and Amara?

(A) $\frac{2}{7}$

(B) $\frac{1}{21}$

(C) $\frac{2}{21}$

(D) $\frac{1}{42}$

(E) $\frac{13}{42}$

36. MacCracken Associates is a management consulting company whose talent pool includes accountants, lawyers, and PhDs as follows:

32 are accountants

28 are lawyers

22 are PhDs

9 are both accountants and lawyers

8 are both accountants and PhDs

6 are both lawyers and PhDs

4 have all three of these credentials

17 have none of these three credentials

How many individuals make up the talent pool at MacCracken Associates?

(F) 76

(G) 80

(H) 82

(J) 99

(K) 126

Solutions to Practice Problems

So, how did you do? Find out here, where I provide the answers to the 36 practice questions from the preceding section, complete with the worked-out solutions.

1. E. Begin by noticing that 6 is half of 12, 9 is half of 18, and so on. Then, you need to find out why 12 goes to 9, 18 goes to 15, and so on. As you probably noticed, subtracting 3 is the answer. So the sequence is generated by an alternating rule: Multiply by 2, subtract 3, multiply by 2, subtract 3, and so on. Thus, the sequence continues as follows:

$$6, 12, 9, 18, 15, 30, 27, 54 \ldots$$

The eighth number in the sequence is 54, so the correct answer is Choice (E).

2. **G.** Danielle started the day with $20. In the morning, she earned $(\$10 \times 4) + \$5 = \$45$, so she had $65. In the afternoon, she earned $\$12 \times 3 = \$36$, giving her a total of $101. The cab ride cost Danielle $7, so she ended the day with $\$101 - \$7 = \$94$.

3. **B.** The graph shows that the five people have sold 16, 12, 19, 14, and 10 books of tickets, so you can plug these numbers into the formula for the mean:

$$\text{Mean} = \frac{\text{Sun of values}}{\text{Number of values}} = \frac{16+12+19+14+10}{5} = \frac{71}{5} = 14.2$$

4. **G.** Begin by distributing to remove the parentheses:

$$\begin{aligned} 5(x-2) &= 3(2x-6) \\ 5x-10 &= 6x-18 \end{aligned}$$

Isolate x and solve:

$$\begin{aligned} -10 &= x-18 \\ 8 &= x \end{aligned}$$

5. **E.** Begin by plugging in –2 for k and 3 for n:

$$n^4 + 5kn - k^3 = 3^4 + 5(-2)(3) - (-2)^3$$

Evaluate the expression using the order of operations: Start with the powers, then do multiplication, and finally work any addition and subtraction. Here's what the calculations look like:

$$= (3)(3)(3)(3) + 5(-2)(3) - (-2)(-2)(-2) = 81 - 30 + 8 = 59$$

6. **G.** Let x be the number. Then you can make the following equation:

$$5(x-1) = 4x + 0.5$$

Simplify and solve:

$$\begin{aligned} 5x-5 &= 4x+0.5 \\ x-5 &= 0.5 \\ x &= 5.5 \end{aligned}$$

7. **A.** Cross-multiply to remove the fractions:

$$\begin{aligned} \frac{n+4}{9} &= \frac{6-n}{11} \\ 11(n+4) &= 9(6-n) \end{aligned}$$

Distribute on both sides of the equation and simplify:

$$\begin{aligned} 11n+44 &= 54-9n \\ 20n+44 &= 54 \\ 20n &= 10 \\ n &= \frac{1}{2} \end{aligned}$$

Thus, substitute $\frac{1}{2}$ for n in $13n - 2n^2$:

$$13\left(\frac{1}{2}\right) - 2\left(\frac{1}{2}\right)^2 = \frac{13}{2} - \frac{1}{2} = \frac{12}{2} = 6$$

8. **J.** First isolate the x term on one side of the equation:

$$ax + by = c$$
$$ax = c - by$$

Next divide both sides of the equation by *a*:

$$x = \frac{c - by}{a}$$

9. **B.** Simplify by factoring both the numerator (top number) and denominator (bottom number) and then canceling out the common factor $j + k$:

$$\frac{10j + 10k}{j^2 - k^2} = \frac{10(j+k)}{(j+k)(j-k)} = \frac{10}{j-k}$$

Substitute 100 for $j - k$ and simplify:

$$= \frac{10}{100} = 0.1$$

10. **J.** Kwame packed 4 suits, 6 shirts, and 12 ties, and he can wear any combination of those. Multiplying these three numbers gives you $4 \times 6 \times 12 = 288$.

11. **C.** To remove the fractions from the equation $\frac{n}{10} - \frac{1-n}{2} = \frac{n+3}{5}$, multiply each term by the common denominator of 10:

$$(10)\frac{n}{10} - (10)\frac{1-n}{2} = (10)\frac{n+3}{5}$$

Simplify by dividing 10 by the denominator and then multiplying by the numerator:

$$1(n) - 5(1-n) = 2(n+3)$$
$$n - 5 + 5n = 2n + 6$$

Solve for *n*:

$$4n = 11$$
$$n = \frac{11}{4}$$

12. **H.** To begin, you need to find out how many people participated in the poll: $225 + 185 + 175 + 90 + 75 = 750$. Exactly 90 of these people have 3 pets, and 75 have more than 3. So 165 people have 3 or more pets. To find the percentage, make a fraction as follows:

$$\frac{165}{750} = 0.22 = 22\%$$

13. **C.** Let x equal the number, and then translate the words into the following equation:

$$2x - 6 = -4(x + 18)$$

Solve for *x*:

$$2x - 6 = -4x - 72$$
$$6x - 6 = -72$$
$$6x = -66$$
$$x = -11$$

Thus, when you add 14 to the number, the result is $-11 + 14 = 3$.

14. K. To find the value of m, you need to express both sides of the equation in terms of the same base. The base 3 works well because:

$$3^2 = 9 \quad \text{and} \quad 3^{\frac{1}{2}} = \sqrt{3}$$

Thus, you can change the equation as follows:

$$9^{m-3} = \sqrt{3}$$
$$\left(3^2\right)^{m-3} = 3^{\frac{1}{2}}$$
$$3^{2(m-3)} = 3^{\frac{1}{2}}$$

Because the bases are equal, the exponents must be equal. As a result, you can drop the bases and solve for m:

$$2(m-3) = \frac{1}{2}$$
$$2m - 6 = \frac{1}{2}$$
$$2m = \frac{1}{2} + 6$$
$$2m = \frac{13}{2}$$
$$m = \frac{13}{4}$$

15. A. Multiply both sides of the equation by $2x - 9$ to get rid of the fraction:

$$\frac{4x - 3y}{2x - 9} = 2$$
$$4x - 3y = 2(2x - 9)$$

Distribute on the right side of the equation:

$$4x - 3y = 4x - 18$$

Subtract $4x$ from both sides:

$$-3y = -18$$

The x terms drop out, allowing you to solve for y:

$$y = 6$$

16. J. Remove the absolute value bars by separating the equation $|17 - 2p| = 25$ into two equations:

$$17 - 2p = 25 \qquad\qquad 17 - 2p = 25$$
$$-2p = 8 \qquad\qquad -2p = -42$$
$$p = -4 \qquad\qquad p = 21$$

The only positive value of p is 21, so the correct answer is Choice (J).

17. E. Begin by writing down 1 and 84 with space between them:

1 84

The number 84 is even as noted by $84 \div 2 = 42$. So put these two numbers at the beginning and end of the list:

1 2 42 84

The number 84 is divisible by 3 ($84 \div 3 = 28$), so add these two numbers to the list:

1 2 3 24 42 84

Continue in this fashion until you reach the middle factors:

1 2 3 4 6 7 12 14 21 24 42 84

As you can see, 84 has exactly 12 factors, so the correct answer is Choice (E).

18. H. The greatest factor among 8, 12, and 20 is 4, because it divides all three of these numbers evenly; therefore, you can rule out Choices **(F)** and **(G)**. The lowest exponent among the three x's is 1, so you can rule out Choice **(K)**. The lowest exponent among the three y's is 2, so you can rule out Choice **(J)**. Therefore, the correct answer is Choice **(H)**.

19. E. The scores are 78, 83, 83, 84, 88, 91, 93, 93, 93, 95, and 96. The median is the middle value among these scores, so it's 91. The mode is the most common value, so it's 93. The median is two points lower than the mode, so the correct answer is Choice (E).

20. J. Let m be the marked-down price of the TV before the 10 percent tax was included. Thus, the price of the TV *with* tax included was 110 percent of m, and this amount was \$880. So

$$\begin{aligned}(110\%)(m) &= 880\\ 1.1m &= 880\\ m &= \frac{880}{1.1}\\ m &= 800\end{aligned}$$

As you can see, the marked-down price before tax was \$800. This amount was 20 percent off of the original price, so it represents 80 percent of the original price. Let p be the original price. So

$$\begin{aligned}(80\%)(p) &= 880\\ 0.8p &= 880\\ p &= \frac{880}{0.8}\\ p &= 1000\end{aligned}$$

21. D. Begin by listing the multiples of 6, 8, and 9:

Multiples of 6: 6, 12, 18, 24, 30, 36, 42, 48, 54, 60, 66, **72** . . .
Multiples of 8: 8, 16, 24, 32, 40, 48, 56, 64, **72**, 80 . . .
Multiples of 9: 9, 18, 27, 36, 45, 54, 63, **72**, 81, 90 . . .

Because $m = 72$, you know that the correct answer is Choice (D).

22. **G.** Begin by splitting the equation into two separate equations:

$$7n+1=3n-11 \qquad 7n+1=-(3n-11)$$

Solve both equations:

$$\begin{aligned} 4n+1&=-11 \\ 4n&=-12 \\ n&=-3 \end{aligned} \qquad \begin{aligned} 7n+1&=-3n+11 \\ 10n+1&=11 \\ 10n&=10 \\ n&=1 \end{aligned}$$

$-3\times 1=-3$, so the correct answer is Choice (**G**).

23. **C.** The school has 15 teachers and a 2:3 ratio of administrators to teachers, so make a proportion like this:

$$\frac{\text{Administrators}}{\text{Teachers}}=\frac{2}{3}$$

Plug in 15 for teachers and a for administrators, and then cross-multiply and solve for a:

$$\begin{aligned} \frac{a}{15}&=\frac{2}{3} \\ 3a&=30 \\ a&=10 \end{aligned}$$

The school has 10 administrators and a 2:35 ratio of administrators to students, so create the following proportion:

$$\frac{\text{Administrators}}{\text{Students}}=\frac{2}{35}$$

Plug in 10 for administrators and s for students, and then cross-multiply and solve for s:

$$\begin{aligned} \frac{10}{s}&=\frac{2}{35} \\ 2s&=350 \\ s&=175 \end{aligned}$$

The school has 175 students, 10 administrators, and 15 teachers, so the total of all these is $175+10+15=200$.

24. **F.** Let x equal the number, and then create the following equations:

$$\frac{x}{3}+13=2(x-36)$$

To eliminate the fraction, multiply every term by 3:

$$x+39=6(x-36)$$

Simplify and solve for x:

$$\begin{aligned} x+39&=6x-216 \\ 39&=5x-216 \\ 255&=5x \\ 51&=x \end{aligned}$$

25. **D.** According to the graph, Lakshmi sold 2 houses in January, 3 in February, and 5 in March, so she sold a total of 10 during those three months.

26. **G.** Lakshmi sold 2 houses in January, 3 in February, 5 in March, 6 in April, 7 in May, and 4 in June. Place this information into the formula for the mean to get your answer:

$$\text{Mean} = \frac{\text{Sum of values}}{\text{Number of values}} = \frac{2+3+5+6+7+4}{6} = \frac{27}{6} = 4.5$$

27. **B.** To isolate the square root expression, add x to both sides of the equation:

$$\sqrt{x^2+1} - x = 2$$
$$\sqrt{x^2+1} = x+2$$

Square both sides of the equation:

$$\left(\sqrt{x^2+1}\right)^2 = (x+2)^2$$

Simplify both sides:

$$x^2+1 = (x+2)(x+2)$$
$$x^2+1 = x^2+4x+4$$

Subtract x^2 from both sides and solve for x:

$$1 = 4x+4$$
$$-3 = 4x$$
$$-\frac{3}{4} = x$$

28. **J.** Begin by counting the number of ways each coin can land as independent events. Because each coin has 2 sides, each one can land 2 different ways as independent events.

Multiply to find the number of possible outcomes: $2 \times 2 \times 2 \times 2 = 16$. Of these, count the number of ways in which exactly 2 coins come up heads:

Penny and Nickel	Penny and Dime	Penny and Quarter
	Nickel and Dime	Nickel and Quarter
		Dime and Quarter

Thus, 6 outcomes are the target outcome. Plug these numbers into the formula for probability:

$$\text{Probability} = \frac{\text{Target outcomes}}{\text{Total outcomes}} = \frac{6}{16} = \frac{3}{8}$$

29. **B.** To answer the question, isolate the expression $4t - u$ on one side of the equation. Begin by multiplying both sides by $2t - 1$ and simplifying:

$$\frac{2t-3u}{1-2t}=5$$
$$2t-3u=5(1-2t)$$
$$2t-3u=5-10t$$
$$12t-3u=5$$

Factor out a 3 on the left side:

$$3(4t-u)=5$$

Now divide both sides of the equation by 3:

$$4t-u=\frac{5}{3}$$

30. **K.** Let x be the number of points Jerome needs to score in his fifth game to get an average of 20 points over five games. Plug this information into the formula for an average:

$$\text{Mean}=\frac{\text{Sum of values}}{\text{Number of values}}$$
$$20=\frac{13+17+19+21+x}{5}$$

Simplify and solve for x:

$$20=\frac{70+x}{5}$$
$$100=70+x$$
$$30=x$$

31. **D.** Let the number of sculptors be represented by the variable x. Then the number of painters is $2x$. The number of students who are both painters and sculptors is 15, and the number of students who are neither painters nor sculptors is 12. The following Venn diagram encapsulates this information:

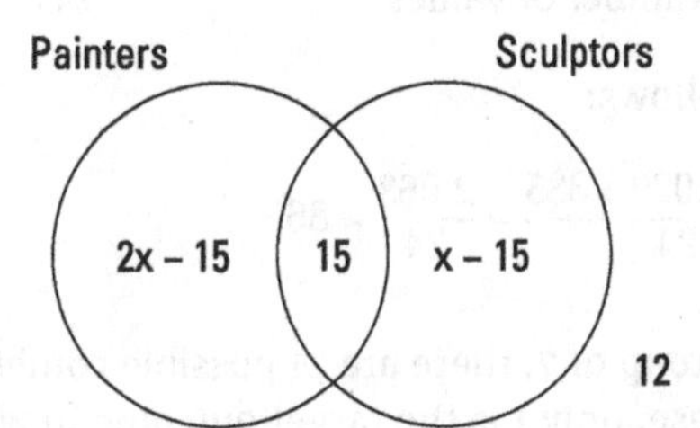

Thus, the $2x - 15$ students are painters but not sculptors, and $x - 15$ students are sculptors but not painters. There are 75 students altogether, so you can make and solve the following equation:

$$\begin{aligned} 2x - 15 + 15 + x - 15 + 12 &= 75 \\ 3x - 3 &= 75 \\ 3x &= 78 \\ x &= 26 \end{aligned}$$

Thus, the number of painters who are not sculptors is as follows:

$$2x - 15 = 2(26) - 15 = 52 - 15 = 37$$

32. J. Of the 86 people who work at the Wallace Street location, 61 are union members. Plug these values into the formula for probability as follows:

$$\text{Probability} = \frac{\text{Target outcomes}}{\text{Total outcomes}} = \frac{61}{86} \approx 0.709 \approx 71\%$$

33. E. The graph provides the following breakdown by percentage of the four main subgroups:

$$\text{Branch Avenue management} = \frac{18}{151} \approx 12\%$$

$$\text{Branch Avenue union members} = \frac{47}{151} \approx 31\%$$

$$\text{Wallace Street management} = \frac{25}{151} \approx 17\%$$

$$\text{Wallace Street union memebers} = \frac{61}{151} \approx 40\%$$

Thus, the two subgroups that total 17% and 31% are the Branch Avenue union members and the Wallace Street management.

34. K. Find the weighted mean by multiplying each of the three average exam scores by the number of students in the group that received that score, and then dividing by 24:

$$\text{Mean} = \frac{\text{Sum of values}}{\text{Number of values}} = \frac{(94 \times 7) + (85 \times 12) + (77 \times 5)}{24}$$

Simplify as follows:

$$= \frac{658 + 1{,}020 + 385}{24} = \frac{2{,}063}{24} \approx 86$$

35. B. Among a group of 7, there are 21 possible combinations of 2 students who might be chosen. Of these, only 1 is the target outcome in which both Erin and Amara are selected. Thus:

$$\text{Probability} = \frac{\text{Target outcomes}}{\text{Total outcomes}} = \frac{1}{21}$$

36. G. Begin by making a Venn diagram with three circles.

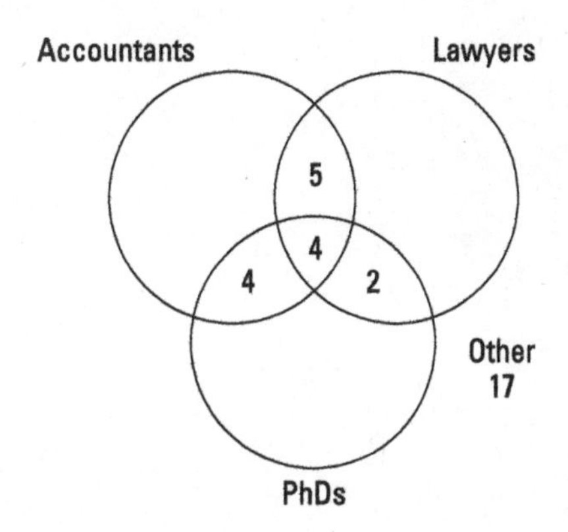

Notice that to start, I filled the 4 students who have all three credentials and the 17 students who have none of these credentials. Next, I filled in the groups of 9, 8, and 6 people who have at least two credentials.

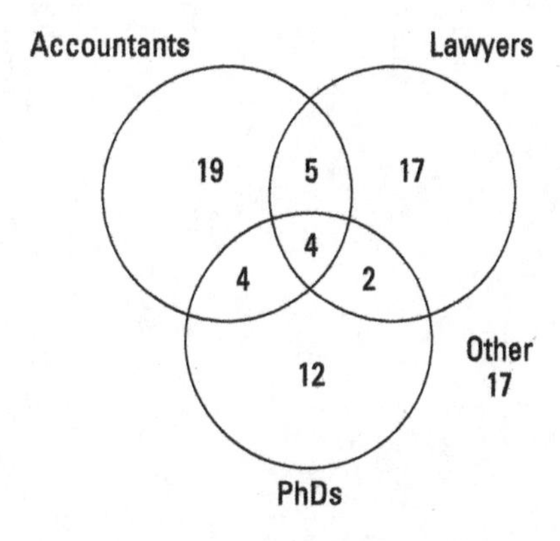

To finish up, complete the chart as shown, so that the totals of accountants, lawyers, and PhDs equal, 32, 28, and 22 respectively. To answer the question, add all the values in the Venn diagram:

$$19 + 17 + 12 + 5 + 4 + 2 + 4 + 17 = 80$$

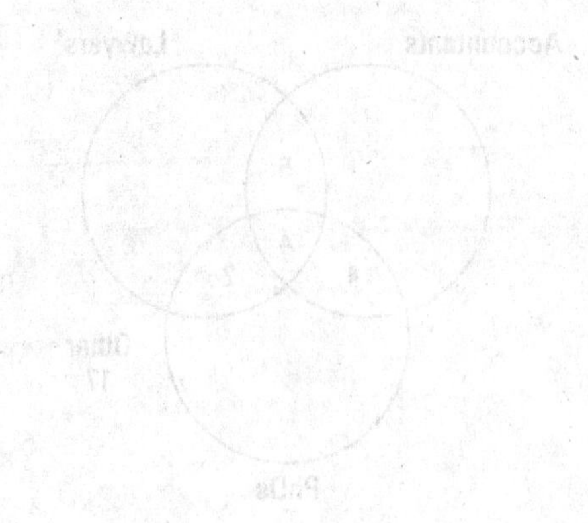

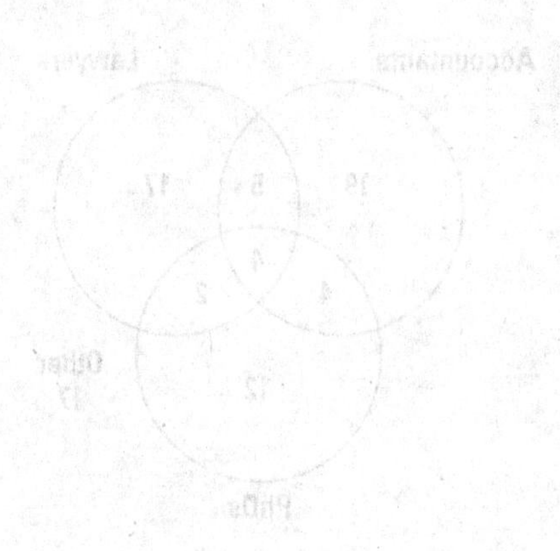

3 Digging In to Intermediate Algebra and Coordinate Geometry

IN THIS PART . . .

Solving inequalities, systems of equations, and quadratic equations, and understanding direct and indirect proportions and function notation.

Working on the *xy*-graph with linear functions, quadratic functions, higher order polynomials, transformations of functions, and circles.

Solving practice problems for all of these topics.

IN THIS CHAPTER

» Working with inequalities

» Using systems of equations

» Reviewing direct and inverse proportions

» Solving quadratic equations

» Understanding functions and function notation

Chapter 8 Moving to Intermediate Algebra

Intermediate algebra takes your basic algebra skills (discussed in Chapter 5) to the next level. In this chapter, you review the subtleties and complexities of algebra that are normally covered at the end of Algebra I and in Algebra II.

You begin by working with inequalities such as >, <, ≥, and ≤, including inequalities that contain absolute value. Then you focus on systems of equations, which are sets of equations with more than one variable that must be solved as a group. Systems of equations are often useful for solving word problems. Next, you see how to work with pairs of variables that are either directly proportional or inversely proportional.

You continue with a look at solving all-important quadratic equations, using both factoring and the quadratic formula. The chapter wraps up with an introduction to functions, including modeling with functions, functional notation, and finding the domain and range of a function.

Knowing More or Less about Inequalities

An *inequality* is a mathematical statement that tells you two values are unequal. Most of the inequalities you'll work with on the ACT are of four types, which I show in Table 8-1.

Inequalities differ from algebraic equations because usually you can't solve for the actual value of the variable. Instead, inequalities are solved for a *solution set,* which is a set of values that all make the inequality true. A solution set itself is usually expressed as a much simpler version of the original inequality.

TABLE 8-1: Four Types of Inequalities

Direction	Exclusive (not equal to)	Inclusive (equal to)
Greater than	>	≥
Less than	<	≤

An ACT question may ask you for the solution set or for an equivalent inequality, both of which are essentially the same thing. A question also may ask you to identify which of five values is in the solution set or, conversely, which is *not* in the solution set.

TIP

Before you begin solving an inequality, notice whether it's exclusive (< or >) or inclusive (≥ or ≤). The process of solving an inequality may change the *direction* of the inequality but not whether it's exclusive or inclusive. This fact often allows you to easily rule out one or more wrong answers.

In this section, I show you how to use algebra to solve a variety of inequalities, including those that contain absolute value. (For a refresher on absolute value, see Chapter 5.)

Becoming a pro at inequalities

Solving inequalities is virtually identical to solving equations, with one important difference: When you multiply or divide by a negative number, you must reverse the direction of the inequality.

EXAMPLE

Which of the following is the solution set of the inequality $8 - 5p > 23$?

(A) $p > 3$

(B) $p < 3$

(C) $p \geq 3$

(D) $p > -3$

(E) $p < -3$

To begin, notice that Choice (C) changes the inequality from exclusive (>) to inclusive (≥), so you can rule out this answer immediately. Now your goal is to isolate *p*, just as you would solve an equation. Begin by subtracting 8 from both sides:

$$8 - 5p > 23$$
$$-5p > 15$$

Your next step is to divide both sides by –5. As you do this, reverse the direction of >, changing it to <:

$$p < -3$$

So the correct answer is Choice (E).

WARNING

When working with inequalities, remember that the *only* time you need to reverse the direction of the inequality is when the actual step you're taking is multiplication or division by a negative number. This next example shows how confusion can arise and how to avoid it.

EXAMPLE

Which of the following is equivalent to the inequality $-4(1-a) < 2a - 8$?

(F) $a < 0$

(G) $a < -1$

(H) $a > -1$

(J) $a < -2$

(K) $a > -2$

To answer this question, isolate a as usual. Begin by distributing –4 on the left side of the inequality. Although this step involves multiplication by a negative number, you are *not* multiplying the entire inequality by a negative, so *don't* reverse the direction of the inequality:

$$-4(1-a) < 2a - 8$$
$$-4 + 4a < 2a - 8$$

Next, subtract $2a$ from both sides (again, no need to reverse):

$$-4 + 2a < -8$$

Now add 4 to both sides:

$$2a < -4$$

Finally, divide both sides by 2. Again, even though the right side of the inequality is negative, the actual step you're taking does *not* involve multiplying or dividing by a negative number, so *don't* reverse the inequality:

$$a < -2$$

Thus, the correct answer is Choice (J).

WARNING

Be careful when solving inequalities that have fractions. To simplify fractions out of an equation, you may need to cross-multiply. When working with inequalities, move any minus signs in front of a fraction or in the denominator (bottom number) up to the numerator (top number) so you can avoid multiplying or dividing by a negative number.

EXAMPLE

Which of the following is equivalent to $-\frac{n}{2} \leq \frac{n+1}{3}$?

(A) $n < \frac{2}{5}$

(B) $n \leq \frac{2}{5}$

(C) $n \geq \frac{2}{5}$

(D) $n \leq -\frac{2}{5}$

(E) $n \geq -\frac{2}{5}$

For starters, the inequality is ≤, so you can rule out any answer that includes either < or >; therefore, Choice (A) is wrong. To simplify this inequality, you need to cross-multiply, which involves multiplying by the denominator. To avoid the confusion of multiplying by a negative number, begin by moving the minus sign on the left side of the inequality to the numerator:

$$\frac{-n}{2} \leq \frac{n+1}{3}$$

This adjustment doesn't change the fraction, but it makes it safer to work with, because you don't have to multiply by a negative number. Now you can cross-multiply as usual, and then solve for *n:*

$$-3n \le 2(n+1)$$
$$-3n \le 2n+2$$
$$-5n \le 2$$

At this point, you need to divide both sides by –5, so reverse the inequality as usual:

$$n \ge -\frac{2}{5}$$

Therefore, the correct answer is Choice (E).

Separate but unequal: Solving inequalities with absolute value

As when solving an equation that includes an expression with absolute value, you also need to split an inequality with absolute value into two separate inequalities. However, keep in mind one twist: One of the two resulting inequalities is simply the original inequality with the bars removed. The other inequality is the original inequality with

- The bars removed
- The opposite side negated (as with absolute value equations)
- The inequality reversed (as with inequalities when you multiply or divide by a negative number)

REMEMBER

These rules aren't difficult, but they're a little complicated, so be careful to do all three parts correctly.

EXAMPLE

Which of the following values is in the solution set of $|4t-1|<5$?

(A) 0

(B) 2

(C) –2

(D) 4

(E) –4

Begin by splitting the inequality:

$$4t-1<5 \qquad 4t-1>-5$$

Notice that the second of these two inequalities has the bars removed, the right side negated, and the inequality sign reversed. You're now ready to solve both of these inequalities for *t:*

$$4t<64 \qquad 4t>-4$$
$$t<1.5 \qquad t>-1$$

To make these inequalities a little easier to read, put them in the following form:

$$-1<t<1.5$$

Thus, 0 falls into the range of solutions, so the right answer is Choice (A).

In some cases, the solution to an inequality with absolute value can lead to a pair of inequalities that appear to contradict each other. When this happens, both inequalities aren't true, but at least one of them is, so link them with the word *or*. This concept is a little tricky, so don't worry if it's not making sense. The next problem provides a concrete example.

EXAMPLE

What is the solution set for $\frac{|2n-5|+7}{5} \geq 2$?

(F) $n \leq 4$

(G) $1 < n < 4$

(H) $1 \leq n \leq 4$

(J) $n < 1$ or $n > 4$

(K) $n \leq 1$ or $n \geq 4$

Before you begin, notice that the original inequality is ≥, so no solution can include either < or >. As a result, you can rule out Choices (G) and (J). Now isolate $|2n-5|$ on the left side of the inequality:

$$|2n-5|+7 \geq 10$$
$$|2n-5| \geq 3$$

You're now ready to remove the bars and split the inequality:

$$2n-5 \geq 3 \qquad 2n-5 \leq -3$$

Notice that the second of these two inequalities has the bars removed, the right side negated, and the inequality sign reversed. You're now ready to solve the first one:

$$2n-5 \geq 3$$
$$2n \geq 8$$
$$n \geq 4$$

Next, solve the second inequality:

$$2n-5 \leq -3$$
$$2n \leq 2$$
$$n \leq 1$$

Notice that the two solutions ($n \geq 4$ and $n \leq 1$) seem to contradict each other: If n is greater than 4, how can it be less than 1? When this situation occurs, either solution can be true, so link the two resulting solutions with the word *or*:

$n \leq 1$ or $n \geq 4$

Thus, the correct answer is Choice (K).

WARNING

Be extra careful when working with an inequality that sets an absolute value either *greater than* (>) or *greater than or equal to* (≥) another value that includes a variable. This type of inequality can sometimes produce a *false* (or *extraneous*) *solution* — that is, a solution that appears correct but doesn't work when plugged back into the problem. The next example shows you how and why this can happen.

EXAMPLE

Which of the following is the solution set for $|x-3|>2x$?

(A) $x<1$

(B) $x<0$

(C) $x<-1$

(D) $x<-2$

(E) $x<-3$

To begin, remove the absolute value bars, split the inequality, and solve each separately:

$$\begin{array}{ll} x-3>2x & x-3<-2x \\ -3>x & -3<-3x \\ & 1>x \end{array}$$

According to this result, $x<1$ *and* $x<-3$ both appear correct, so you may be tempted to choose Choice (E). However, if this answer were correct, then $x=0$ should be outside the solution set. So plugging 0 into the original inequality should give you the wrong answer:

$$\begin{aligned} &|x-3|>2x \\ &|0-3|>2(0) \\ &3>0 \quad \text{Correct!} \end{aligned}$$

This solution is unexpected. In fact, $x=0$ is in the solution set for this inequality.

What went wrong? Take another look at the original inequality:

$$|x-3|>2x$$

This inequality sets an absolute value *greater than* 2x. So if x is *any* negative number, the absolute value (which can never be negative) must be in the solution set. Therefore, the solution $x<-3$ is false because it tells you that only certain negative values of x are in the solution set. Throwing out this false solution leaves you with the correct answer, which is $x<1$; so the correct answer is Choice (A).

Beating the System: Dealing with Systems of Equations

Much of elementary algebra focuses on solving equations that have a single variable. In contrast, in intermediate algebra you're bound to come across problems that include more than one variable. A *system of equations* is a set of two or more equations that include two or more variables. Systems of equations can be useful for answering tricky word problems. In this section, I show you how and when to use systems of equations.

Solving systems of equations systematically

To solve a system of equations, you need one equation for every variable in the system. In an ACT question, this usually means two equations and two variables.

You can solve a system of linear equations in two ways:

» **Substitution:** With this technique, you solve one equation for a variable in terms of the other(s), and then you substitute this value into the second equation.

» **Elimination:** To use this method, you add or subtract the two equations in such a way that one variable drops out of the resulting equation.

REMEMBER

Both of these methods are similar in that they allow you to write a single equation in one variable, which you can then solve using your usual bag of algebra tricks. After you know the value of one variable, you can substitute this value back into one of the original two equations (usually the easier one) to get the value of the remaining variable.

Finding values with substitution

Substitution is easier to use when a variable in one equation is already isolated or when it can be isolated easily.

EXAMPLE

If $x+9=y$ and $7x-2=2y$, what is the value of xy?

(A) 48

(B) 49

(C) 50

(D) 51

(E) 52

This question gives you two equations in two variables. In the first equation, y is already isolated on one side of the equation, so substitution should work well. Substitute $x+9$ for y in the second equation:

$$7x-2=2y$$
$$7x-2=2(x+9)$$

Simplify and solve:

$$7x-2=2x+18$$
$$5x=20$$
$$x=4$$

Now that you know the value of x, substitute this value back into the equation that looks easiest to work with — in this case, the first equation — and solve for y:

$$x+9=y$$
$$4+9=y$$
$$13=y$$

Thus, $x=4$ and $y=13$, so $xy=52$. The correct answer is Choice (E).

Solving systems using elimination

The technique of elimination is easier to use when both equations contain essentially the same term. Check out the following example to see what I mean.

EXAMPLE

If $4s+5t=9$ and $9s+5t=-11$, what is the value of $s+t$?

(A) 2

(B) 1

(C) 0

(D) –1

(E) –2

Answering this question using substitution would be difficult because neither variable is very easy to isolate on one side of the equations. However, both equations include the term 5*t*, so you can combine the two equations using subtraction.

$$\begin{aligned} 4s+5t &= 9 \\ -(9s)+5t &= -11 \\ \hline -5s &= 20 \end{aligned}$$

When you subtract one equation from the other, the *t* term drops out. The resulting equation is easy to solve:

$$\begin{aligned} -5s &= 20 \\ s &= -4 \end{aligned}$$

As always, when you know the value of one variable, you can substitute this value back into either equation — whichever looks easiest — and solve for the other variable, like this:

$$\begin{aligned} 4s+5t &= 9 \\ 4(-4)+5t &= 9 \\ -16+5t &= 9 \\ 5t &= 25 \\ t &= 5 \end{aligned}$$

So $s=-4$ and $t=5$, meaning $s+t=1$. As a result, the correct answer is Choice (B).

Sometimes before you can do elimination, you may need to multiply one or both of them by a certain number to get two terms to match.

EXAMPLE

If $7x+2y=-1$ and $9x-3y=1.5$, which of the following is the value of y?

(F) 0

(G) –0.25

(H) –0.5

(J) –0.75

(K) –1.25

This system would be difficult to solve using substitution. And to eliminate a variable, you need to make two terms look similar. The *y* terms have smaller coefficients (2 and –3), so they should be easier to work with. You want to multiply both of these equations by different numbers so the coefficients of the *y* terms become the same (disregarding their signs.) I choose the numbers 3 and 2, for reasons that will make sense in a moment:

$$\begin{aligned} 3(7x+2y=-1) &\quad \text{becomes} \quad 21x+6y=-3 \\ 2(9x-3y=1.5) &\quad \text{becomes} \quad 18x-6y=3 \end{aligned}$$

Notice that the two resulting equations have y terms that look very similar — the coefficients are identical except for their sign. If you add these two equations, these similar terms drop out:

$$\begin{array}{r} 21x + 6y = -3 \\ +\ 18x - 6y = \ \ 3 \\ \hline 39x = \ \ 0 \end{array}$$

The equation is now easy to solve:

$$\begin{aligned} 39x &= 0 \\ x &= 0 \end{aligned}$$

Now you can substitute 0 for x in either equation:

$$\begin{aligned} 7x + 2y &= -1 \\ 7(0) + 2y &= -1 \\ 2y &= -1 \\ y &= -0.5 \end{aligned}$$

Thus, the correct answer is Choice (H).

Working word problems using a system of equations

Solving word problems is one of the most common reasons to use a system of equations. Some word problems that would be difficult to approach using a single variable are relatively easy when you use more than one variable. This section shows you what I mean.

Dorian and Micah have been saving money from their summer jobs. If Dorian had twice as much money and Micah had half as much, together they would have \$2,075. And if Micah had twice as much money and Dorian had half as much, together they would have \$2,300. How much money does Dorian have?

(A) \$800

(B) \$850

(C) \$900

(D) \$950

(E) \$1,000

You could solve this problem using only one variable, but that approach would be tricky and would likely lead to a mistake along the way. Instead, use two variables, letting d equal Dorian's money and m equal Micah's money. Set up two equations as follows:

$$\begin{aligned} 2d + \frac{m}{2} &= 2{,}075 \\ 2m + \frac{d}{2} &= 2{,}300 \end{aligned}$$

To eliminate the fractions, multiply both of these equations by 2:

$$\begin{aligned} 4d + m &= 4{,}150 \\ 4m + d &= 4{,}600 \end{aligned}$$

This system of equations is easy to solve using substitution. Begin by isolating m in the first equation:

$$m = 4{,}150 - 4d$$

Now substitute $4{,}150 - 4d$ for m in the second equation, and then solve for d:

$$\begin{aligned} 4(4{,}150 - 4d) + d &= 4{,}600 \\ 16{,}600 - 16d + d &= 4{,}600 \\ 16{,}600 - 15d &= 4{,}600 \\ -15d &= -12{,}000 \\ d &= 800 \end{aligned}$$

Dorian has $800, so the correct answer is Choice (A).

To solve some complicated problems, the only sensible approach is to use several different variables to set up a system of equations and then solve it.

EXAMPLE

Andie, Candie, and Sandie are all selling boxes of greeting cards to help pay for new soccer equipment for their team. On Friday night, Andie had sold as many boxes as Candie and Sandie combined. Over the weekend, each girl sold 10 boxes, so at that point Andie had sold twice as many boxes as Candie. Then on Monday, Andie and Sandie sold 5 more boxes each, bringing the final total for the three girls to 120 boxes. How many boxes did Candie sell?

(F) 10

(G) 12

(H) 15

(J) 16

(K) 20

TIP

This problem is a long and complicated one, and you may do well to skip over it to save time. If you decide to solve it, use variables and build a system of three equations.

Let a, c, and s stand for the amounts that the three girls had sold before the weekend. This problem is so complicated that a chart is helpful. Set up a chart showing how many boxes each girl had sold at the three different times:

Friday	*Weekend*	*Monday*
a	$a + 10$	$a + 15$
c	$c + 10$	$c + 10$
s	$s + 10$	$s + 15$

Now use the chart to begin writing equations. On Friday, Andie had sold twice as many boxes as both Candie and Sandie combined:

$$a = c + s$$

After the weekend, Andie had sold twice as many boxes as Candie:

$$\begin{aligned} a + 10 &= 2(c + 10) \\ a + 10 &= 2c + 20 \\ a &= 2c + 10 \end{aligned}$$

On Monday, the three girls had sold 120 boxes altogether:

$$a + 15 + c + 10 + s + 15 = 120$$
$$a + c + s + 40 = 120$$
$$a + c + s = 80$$

Taken together, here's the resulting system of equations:

$$a = c + s$$
$$a = 2c + 10$$
$$a + c + s = 80$$

To begin solving, substitute $c + s$ for a into the second and third equations:

$$c + s = 2c + 10$$
$$c + s + c + s = 80$$

Simplify both equations:

$$s = c + 10$$
$$2c + 2s = 80$$

Now substitute $c + 10$ for s in the second equation:

$$2c + 2(c + 10) = 80$$
$$2c + 2c + 20 = 80$$
$$4c + 20 = 80$$
$$4c = 60$$
$$c = 15$$

Candie sold 15 boxes, so the correct answer is Choice (H).

Keeping Things in Proportion: Direct and Inverse Proportionality

Proportionality refers to a connection between two variables based on either multiplication or division. Two types of proportionality exist:

- **Direct proportionality:** When a pair of variables is *directly proportional*, the variables tend to rise and fall together. That is, as one increases or decreases, the other does the same.
- **Inverse proportionality:** When a pair of variables is *inversely proportional*, they tend to rise and fall separately. That is, as one increases, the other decreases, and vice versa.

In this section, I show you how to answer a variety of ACT questions that involve direct and inverse proportionality.

Maintaining a balance with direct proportions

Two variables, x and y, are directly proportional when the following equation is true for some constant k:

$$\frac{x}{y} = k$$

In practical terms, direct proportionality simply means that as the value of one variable changes, the other value also must change so that any resulting fraction $\frac{x}{y}$ remains constant.

EXAMPLE

Two variables, a and b, are directly proportional. If $a = 6$, then $b = 18$. Which of the following must be true?

(A) If $a = 1$, then $b = 6$

(B) If $a = 3$, then $b = 9$

(C) If $a = 12$, then $b = 12$

(D) If $a = 18$, then $b = 6$

(E) If $a = 100$, then $b = 200$

The fraction $\frac{a}{b}$ is a constant, and

$$\frac{a}{b} = \frac{6}{18} = \frac{1}{3}$$

Thus, any combination of a and b must make a fraction equivalent to $\frac{1}{3}$. The only such combination is $a = 3$ and $b = 9$, because:

$$\frac{a}{b} = \frac{3}{9} = \frac{1}{3}$$

So the correct answer is Choice (B).

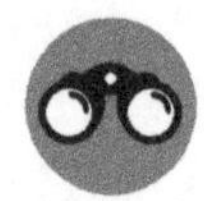

EXAMPLE

Two variables, x and y, are directly proportional such that if $x = 3$, then $y = 5$. What is the value of x when $y = 15$?

(F) 1

(G) 2

(H) 6

(J) 9

(K) 13

When $x = 3$ and $y = 5$:

$$\frac{x}{y} = \frac{3}{5}$$

Thus, any value of $\frac{x}{y}$ must also produce the fraction $\frac{3}{5}$. Substitute 15 for y into the preceding equation:

$$\frac{x}{15} = \frac{3}{5}$$

Cross-multiply and solve for x:

$$5x = 45$$
$$x = 9$$

Therefore, the correct answer is Choice (J).

Turning things around with inverse proportions

Two variables, x and y, are inversely proportional when the following equation is true for some constant k:

$$xy = k$$

Inverse proportionality means that as the value of one variable changes, the other value must also change so that any resulting product xy remains constant.

Two variables p and q are inversely proportional, such that if $p = 4$, then $q = 8$. What is the value of q when $p = 16$?

(A) 1

(B) 2

(C) 4

(D) 16

(E) 32

The product pq is a constant and

$$pq = (4)(8) = 32$$

Thus, $pq = 32$ for all possible pairings of p and q. Substitute 16 for p into this equation:

$$pq = 32$$
$$16q = 32$$
$$q = 2$$

Therefore, the correct answer is Choice (B).

EXAMPLE

If $\frac{t}{u} = \frac{v}{w}$ and $uv = 10$, which of the following must be true?

(F) t and u are inversely proportional

(G) t and v are inversely proportional

(H) t and w are directly proportional

(J) t and w are inversely proportional

(K) u and v are directly proportional

Begin by cross-multiplying:

$$\frac{t}{u} = \frac{v}{w}$$
$$tw = uv$$

Substitute 10 for uv:

$$tw = 10$$

Thus, $tw = k$ for $k = 10$, so t and w are inversely proportional. So the correct answer is Choice (J).

Working with Quadratic Equations and the Roots of Polynomials

A *quadratic equation* is a second-degree polynomial — that is, it has an x^2 term. Here's the standard form of the quadratic equation:

$$ax^2 + bx + c = 0$$

REMEMBER

Quadratic equations differ from the equations you work with in elementary algebra for a variety of reasons. First of all, you can't solve a three-term quadratic equation using the common method of trying to isolate *x*. Instead, two alternative methods for solving quadratic equations are used:

- **Factoring:** In some cases, you can factor one side of a quadratic equation into a pair of linear factors and then split it into two separate equations. This method is usually quick, but it doesn't always work.
- **The quadratic formula:** In every case, you can apply the quadratic formula to solve for *x*. The downside is that even though the formula is effective, it's long and unwieldy to use.

In this section, I show you how and when to use both methods for solving quadratic equations.

TIP

Try factoring first whenever a quadratic equation has relatively small coefficients and an x^2 term with a coefficient of 1 — for example, the equation $x^2 + 9x + 10 = 0$. When a quadratic equation has larger coefficients — and especially when it has an x^2 term with a coefficient that isn't 1 — you may want to use the quadratic formula. This advice goes double if you don't feel confident with the factoring method. And in any case where factoring simply doesn't work, you need to use the quadratic formula.

Factoring to solve quadratic equations

A quick way to solve many simple quadratic equations — especially the ones you're likely to find on the ACT — is *factoring*. When you factor a quadratic equation, you break down the quadratic into the product of two linear terms. Although not every quadratic equation is factorable, many of the problems you run across on the ACT can be solved in this way.

In this section, I first show you how to solve simple factorable equations of the form $x^2 + bx + c = 0$. Then, you see how to solve more difficult equations of the form $ax^2 + bx + c = 0$.

Combining equations when a equals 1

The easier type of factorable quadratic is of the form $x^2 + bx + c = 0$. In other words, it's x^2 term has a coefficient of 1. When working with this type of problem, factor using a pair of numbers that:

- Adds up to *b*
- Multiplies to *c*

The following example shows you how this method of factoring works.

EXAMPLE

If x_1 and x_2 are the two solutions to the quadratic equation $x^2 - 6x + 8 = 0$, with $x_1 < x_2$, then $x_2 - x_1$ equals what?

(A) 0

(B) 2

(C) 6

(D) 7

(E) 9

In the equation $x^2 - 6x + 8 = 0$, the x^2 term has a coefficient of 1, so you can set up a shell for factoring it as follows:

$$(x \quad)(x \quad) = 0$$

To fill in the remaining numbers, you're looking for a pair of integers that adds up to –6 and multiplies to 8. Begin by listing all the pairs of integers that multiply to 8, including those pairs that include negative numbers:

$$1 \times 8 \qquad 2 \times 4$$
$$-1 \times (-8) \qquad -2 \times (-4)$$

Now pick out the pair that adds up to –6:

$$-2 + (-4) = -6$$

Fill these two numbers into the empty shell you made earlier:

$$(x - 2)(x - 4) = 0$$

The good news is that if *either* of these factors equals 0, the whole equation is correct (because 0 multiplied by anything equals 0). So you can split this equation into two separate equations and solve both:

$$x - 2 = 0 \qquad x - 4 = 0$$
$$x = 2 \qquad x = 4$$

The question asks you to subtract the smaller number from the greater one, so $x_2 - x_1 = 4 - 2 = 2$; therefore, the correct answer is Choice (B).

Solving equations when a doesn't equal 1

The more difficult type of factorable quadratic occurs when the leading coefficient *a* isn't equal to 1. In some cases, you may be able to use GCF factoring to get rid of this problem. In other cases, you'll need to use a more complicated method of factoring. The next two questions illustrate these two cases.

EXAMPLE

Which of the following is a positive solution to the equation $3x^2 + 33x - 36 = 0$?

(A) $\frac{1}{3}$

(B) $\frac{2}{3}$

(C) 1

(D) 2

(E) 3

The leading coefficient of $3x^3 + 33x^2 - 36x = 0$ makes this equation look difficult at first. Fortunately, you can factor out a GCF of $3x$:

$$3x^3 + 33x^2 - 36x = 0$$
$$3x\left(x^2 + 11x - 12\right) = 0$$

Now, you can factor the polynomial $x^2 + 11x - 12$ using the method you already know:

$$3x(x+12)(x-1) = 0$$

Set each of these three factors to 0 and solve for x:

$$3x = 0 \qquad x + 12 = 0 \qquad x - 1 = 0$$
$$x = 0 \qquad x = -12 \qquad x = 1$$

Therefore, the only positive solution to this equation is 1, so the correct answer is Choice (C).

This method may not work with a more difficult quadratic equation. Here's an example to show you how to solve a quadratic when you can't factor out the leading coefficient.

EXAMPLE

Which of the following is a positive solution to the equation $3x^2 + 4x - 4 = 0$?

(F) $\frac{1}{3}$

(G) $\frac{2}{3}$

(H) 1

(J) 2

(K) 3

In this equation, notice coefficients of the three terms are as follows:

$$a = 3 \qquad b = 4 \qquad c = -4$$

You can't simplify these values by factoring out a GCF. Instead, you need to split $4x$ into two equivalent terms whose coefficients multiply to $ac = -12$ and add up to $b = 4$. The values that work are 6 and -2:

$$3x^2 + 4x - 4 = 0$$
$$3x^2 + 6x - 2x - 4 = 0$$

Now, factor the left side of this equation by grouping the first pair of terms and the last pair of terms separately:

$$3x(x+2) - 2(x+2) = 0$$

WARNING

Before moving on, make sure that the expressions inside the two sets of parentheses are the same. If they're not, check your work and try to figure out what went wrong!

Assuming these expressions are the same, however, you can now combine like terms to rewrite the left side of the equation as follows:

$$(3x-2)(x+2) = 0$$

Solve for both values of x:

$$3x - 2 = 0 \qquad x + 2 = 0$$
$$3x = 2 \qquad x = -2$$
$$x = \frac{2}{3}$$

Thus, the only positive solution to this equation is $\frac{2}{3}$, so the correct answer is Choice (G).

Choosing the quadratic formula when all else fails

No matter how good you get at factoring to solve quadratic equations, eventually you'll be faced with an equation that isn't factorable. To solve these equations, use the quadratic formula.

The *quadratic formula* gives you the value of x for any quadratic equation of the form $ax^2 + bx + c = 0$, when you input the values of a, b, and c. Here's the quadratic formula:

$$x = \frac{-b \pm \sqrt{b^2 - 4ac}}{2a}$$

This formula produces two values, because the symbol $\pm$ stands for "plus or minus."

TIP

Although the quadratic formula always gives the correct answer, it's complicated and easy to make a mistake with, so use it only when you're clear that factoring doesn't work.

EXAMPLE

What is the sum of the values of x that satisfy the equation $x^2 + 10x + 7 = 0$?

(A) 0

(B) -10

(C) $-5 + 3\sqrt{2}$

(D) $-10 + 6\sqrt{2}$

(E) $-10 - 6\sqrt{2}$

Your first thought may be to factor $x^2 + 10x + 7$, but this proves impossible. After all, only two pair of integers multiply to 7: 1 and 7, and -1 and -7. And, as you can see, neither of these pairs adds up to 10. The only way to find the value of x is to use the quadratic formula, using the values of the coefficients, as follows:

$$a = 1$$
$$b = 10$$
$$c = 7$$

Plug these values into the quadratic formula:

$$x = \frac{-b \pm \sqrt{b^2 - 4ac}}{2a}$$
$$= \frac{-10 \pm \sqrt{(10)^2 - 4(1)(7)}}{2(1)}$$

Now simplify:

$$= \frac{-10 \pm \sqrt{72}}{2}$$
$$= \frac{-10 \pm \sqrt{36}\sqrt{2}}{2}$$
$$= \frac{-10 \pm 6\sqrt{2}}{2}$$
$$= -5 \pm 3\sqrt{2}$$

You actually have two solutions: $x = -5 + 3\sqrt{2}$ and $x = -5 - 3\sqrt{2}$. Adding these two solutions together gives you:

$$-5 + 3\sqrt{2} - 5 - 3\sqrt{2} = -5 - 5 = -10$$

So the correct answer is Choice (B).

Making Connections with Functions

A *function* is a mathematical connection between two numbers: an *input* (or independent variable) and an *output* (or dependent variable). Most commonly, a function links an input of x with an output of y. For example:

$$y = 3x$$

You can draw an input-output table to show how any input value of x determines the output value of y:

x	0	1	2	3	4	5 . . .	10 . . .	100 . . .
y	0	3	6	9	12	15 . . .	30 . . .	300 . . .

One key feature of a function is that each input value of x results in exactly one output value of y. That is, in an input-output chart, the x-values cannot be repeated, but the y-values can.

Often, the y variable is set equal to $f(x)$. For example:

$$f(x) = 3x$$

The use of $f(x)$ to describe a function is called *functional notation*, which you work with later in this chapter. The variables y and $f(x)$ are essentially interchangeable, but y generally is used for graphing specific functions in coordinate geometry (see Chapter 9) and $f(x)$ is more common for practical uses of functions, such as modeling real-world situations.

In this section, I provide a variety of common skills to help you answer ACT function questions quickly and correctly.

Using functions as models

Functions typically are used to *model* the real world — that is, to create an equation that takes the place of a complex table of numbers. A typical situation asks you to model the cost over time of an initial amount of money plus another amount paid at regular intervals.

For example, suppose you open a savings account with $1,000 and begin saving $100 a week. You could use the following table to keep track of how much money you will have saved in any particular week:

x = Week number	0	1	2	3	4	5	6	7
y = Dollars saved	1,000	1,100	1,200	1,300	1,400	1,500	1,600	1,700

This table is useful but limited. For example, if you want to find out how much money you'll have after two years (104 weeks), the table is no longer practical. A more compact way to express this relationship between input value x (the week number) and the output value y (the number of dollars saved) is with the following function:

$$y = 100x + 1{,}000$$

Now just plug in 104 as your input value for x and solve for y:

$$y = 100(104) + 1{,}000 = 10{,}400 + 1{,}000 = 11{,}400$$

ACT questions often ask you to use a function to model a real-world situation like this one and then use it to find specific information. You'll know you've come across one of these questions when you see the following type of setup for a series of questions.

A gym has a sign-up fee of $250 and a monthly charge of $40 per month. Use this information to answer the next two questions.

EXAMPLE

Which of the following functions models the cost of the gym, outputting a dollar amount (y) when you input a number of months (x)?

(A) $y = 40x$

(B) $y = 250x$

(C) $y = 40x + 250$

(D) $y = 40x - 250$

(E) $y = 250x + 40$

When modeling with a function, drawing a quick input-output table like the following can be helpful:

x = Number of months	0	1	2	3	4	5	6	7
y = Number of dollars	250	290	330	370	410	450	490	530

Setting up the table clarifies how to proceed in writing the function. Multiply the number of months (x) by 40 and then add 250:

$$y = 40x + 250$$

So the correct answer is Choice (C).

EXAMPLE

How much money will a 36-month membership cost?

(F) $940

(G) $1,200

(H) $1,440

(J) $1,690

(K) $9,400

Use the function $y = 40x + 250$, which you created in the previous example question, to answer this one. Substitute 36 for x and solve for y:

$$y = 40(36) + 250 = 1{,}440 + 250 = 1{,}690$$

Therefore, the correct answer is Choice (J).

WARNING

Be careful when an initial cost helps to cover one or more payments — you may have to subtract a number from *x* before multiplying. The next example shows you how this type of problem works.

EXAMPLE

A personal storage facility is offering a special deal for new customers: The first three months costs a total of $10, and then each subsequent month costs $50. Which of the following functions models the monthly cost of this facility for all input values of three or more months?

(A) $f(x)=10x+50$

(B) $f(x)=50x+10$

(C) $f(x)=50x+30$

(D) $f(x)=50x-140$

(E) $f(x)=50x-150$

Notice that this example uses *f*(*x*) rather than *y*, but the meaning is the same. In this case, the input variable *x* is the number of months, and the output variable *f*(*x*) is the dollar amount. You may be tempted to write this answer:

$$f(x)=50x+10 \text{ Wrong!}$$

Unfortunately, this answer is too simple, because it doesn't account for the fact that the initial cost of $10 covers the first three months. So you can subtract 3 months from the input value *before* you multiply this number by 50:

$$f(x)=50(x-3)+10$$

Simplify:

$$f(x)=50x-150+10$$

$$f(x)=50x-140$$

This answer makes sense when you think about it: Instead of paying the normal rate of $50 a month for the first three months, which would be $150, you only pay $10 total. So you're saving $140, which you deduct from whatever you end up paying. Therefore, the correct answer is Choice (D).

Defining relationships with functional notation and evaluation

Functional notation is a compact way to define the relationship between an input variable and an output variable. Typically, the input variable in a function is *x* and the output variable is *f*(*x*). For example:

$$f(x)=x^2+1$$

You can substitute any number for *x* and then evaluate the right side of the function as a value. For example, here's how you find the values of *f*(5) and *f*(10):

$$f(5)=5^2+1=25+1=26$$

$$f(10)=10^2+1=100+1=101$$

Thus, $f(5)=26$ and $f(10)=101$.

EXAMPLE

What is the value of $f(8)-f(3)$ for the function $f(x)=x^2-6$?

(A) 55

(B) 58

(C) 64

(D) 72

(E) 73

First find the value of $f(8)$ and $f(3)$:

$$f(8)=8^2-6=64-6=58$$

$$f(3)=3^2-6=9-6=3$$

After you discover that $f(8)=58$ and $f(3)=3$, you have to do the subtraction:

$$f(8)-f(3)=58-3=55$$

Therefore, the correct answer is Choice (A).

In some cases, an ACT question may include more than one function to evaluate. Consider the following example:

EXAMPLE

If $f(x)=5x+4$ and $g(x)=2x^2$, what is the value of $\frac{f(10)}{g(3)}$?

(F) 2

(G) 3

(H) 4

(J) 6

(K) 8

First evaluate $f(10)$ and $g(3)$:

$$f(10)=5(10)+4=54$$

$$g(3)=2x^2=2(3)^2=2(9)=18$$

Now that you know that $f(10)=54$ and $g(3)=18$, you can simplify:

$$\frac{f(10)}{g(3)}=\frac{54}{18}=3$$

So the correct answer is Choice (G).

Inputting variables to functions

When you know how to input numbers into functions and evaluate them, the next step is inputting variables.

In some cases, inputting variables is even easier than inputting numbers because no calculation is necessary. For example, suppose you have the following function:

$$f(x)=3x-5$$

Here's how you input the variable n into the function:

$$f(n)=3n-5$$

In other cases, you may need to simplify the result. For example:

EXAMPLE

What is the value of $f(a)+f(2a+1)$ for the function $f(x)=7x-2$?

(A) $7a+11$

(B) $-7a+3$

(C) $14a+3$

(D) $21a+3$

(E) $-21a+7$

$$f(a)+f(2a+1)=7a-2+7(2a+1)-2$$

Simplify:

$$=7a-2+14a+7-2=21a+3$$

Therefore, the correct answer is Choice (D).

Evaluating compositions of functions

When one function is placed inside another function, the result is a *composition of functions* (also called a *nested function*).

TIP

Compositions of functions look tricky and difficult to understand, but if you evaluate them starting with the *inside function*, they usually turn out to be a lot easier than they appear. For example:

EXAMPLE

If $f(x)=2x+3$ and $g(x)=x^2-5$, then $f(g(4))=$

(A) 12

(B) 15

(C) 16

(D) 20

(E) 25

The first step here is to notice that $g(4)$ is the inside function, so begin by finding this value using $g(x)=x^2-5$:

$$g(4)=4^2-5=11$$

Next, substitute 11 for $g(4)$ into the function $f(g(4))$:

$$f(g(4))=f(11)$$

To complete the problem, use $f(x)=2x+3$ to find the value of $f(11)$:

$$f(11)=2(11)+3=25$$

Therefore, the correct answer is Choice (E).

Finding the inverse of a function

You already know that addition and subtraction are inverse operations, because when you add a number to a value and then subtract out that same number, you're left with the value you started with. Multiplication and division are inverse operations for similar reasons.

Many functions also have inverses. To find the inverse of a function if it exists, follow these three steps:

1. **Change the *f(x)* to *y*.**
2. **Exchange the *x* and *y*-values.**
3. **Solve for *y*.**

Here's an example to show you how it's done:

EXAMPLE

If $f(x) = 5x^3 - 8$, and $f^{-1}(x)$ is its inverse, then $f^{-1}(x) =$

(A) $\sqrt[3]{\frac{x+5}{8}}$

(B) $\sqrt[3]{\frac{x-5}{8}}$

(C) $\sqrt[3]{\frac{x+8}{5}}$

(D) $\sqrt[3]{\frac{x-8}{5}}$

(E) The function has no inverse.

Follow these steps to answer the question:

1. **Change the *f(x)* to *y*:**

   $y = 5x^3 - 8$

2. **Exchange the *x* and *y*-values.**

   $x = 5y^3 - 8$

3. **Solve for *y*.**

$$x + 8 = 5y^3$$

$$\frac{x+8}{5} = y^3$$

$$\sqrt[3]{\frac{x+8}{5}} = y$$

Therefore, the correct answer is Choice (C).

Getting to know domain and range

The *domain* of a function is the set of values that can be inputted into that function (that is, the *x*-values). The *range* of a function is the set of values that the function can output (that is, the *y*-values). In this section, you discover how to find the domain and range for a variety of functions.

Arriving at the domain event

On the ACT, questions involving the domain of a function usually require you to rule out input values of x that would cause problems. The two main problems to watch out for are the following:

- The denominator of a fraction can't equal 0.
- The value inside a square root (radical) can't be a negative number.

If any value of x causes either of these two problems, that value can't be in the domain. When working with fractions, usually only a limited number of values are left out of the domain, as in the following example.

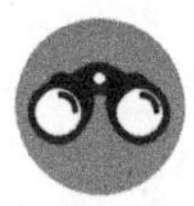

EXAMPLE

Which of the following values of x is NOT in the domain of $f(x) = \frac{x-2}{2x-10}$?

(A) 0

(B) –1

(C) 2

(D) –4

(E) 5

Any value of x that causes the denominator of the fraction to equal 0 can't be in the domain. To find any problematic values, set the denominator equal to 0 and solve for x:

$$2x - 10 = 0$$
$$2x = 10$$
$$x = 5$$

Because 5 can't be in the domain, the correct answer is Choice (E).

The domain of a function must exclude *all* values in which a negative number appears inside a square root. (However, the value 0 inside a square root is acceptable as long as it's not the denominator of a fraction.) Typically, the domain of a function that includes a square root is expressed as an inequality.

EXAMPLE

Which of the following is the domain of the function $f(x) = \sqrt{3x-9}$?

(F) $x > 3$

(G) $x \geq 3$

(H) $x \leq 3$

(J) $x < -3$

(K) $x \geq -3$

The domain of the function can't include a negative number inside the square root. Therefore, the value inside the square root is either greater than or equal to 0. You can set up an inequality that looks like this:

$$3x - 9 \geq 0$$

Now simplify:

$$3x \geq 9$$
$$x \geq 3$$

The correct answer is Choice (G).

Feeling at home with the range

ACT questions involving the range usually focus on functions that input non-negative values. These fall into two main categories:

- **Squares (and even-numbered powers):** For example, no matter what value of x you input, the function $f(x)=x^2$ is never negative. So the range of $f(x)\geq 0$.
- **Absolute value:** For example, the function $f(x)=|x|$ can't be negative either, so the range of $f(x)\geq 0$.

Use these basic principles to think through an ACT question about the range of a function. For example, the range of $f(x)=-x^2$ is $f(x)\leq 0$, because x^2 is always non-negative, and then the minus sign negates the function. Similarly, the range of $f(x)=|x|+3$ is $f(x)\geq 3$, because the minimum value of $|x|$ is 0, and then 3 is added. However, the range of $f(x)=|x+3|$ is $f(x)\geq 0$, because the entire function is an absolute value, whose value can never fall below 0.

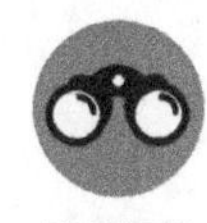

EXAMPLE

Which of the following has a range of $f(x)\geq -2$?

(A) $f(x)=|x-2|$

(B) $f(x)=(x+2)^2$

(C) $f(x)=|x|+2$

(D) $f(x)=|x|-2$

(E) $f(x)=2-|x|$

First, because any function with range of $f(x)\geq -2$ produces some negative values, you can rule out a few answers. A function that's entirely an absolute value can never produce a negative value, so Choice (A) is out. Similarly, a function that's entirely squared can never produce a negative value, so Choice (B) is also ruled out.

The lowest possible value for $|x|$ is 0, so the lowest possible value for both $f(x)=|x|+2$ and $f(x)=-2|x|$ is 2; therefore, you can rule out Choices (C) and (E). By process of elimination, you know that the correct answer is Choice (D). To check, consider this: If $x=0$, then $f(x)=|x|-2=0-2=-2$, and this is the lowest possible value for this function.

IN THIS CHAPTER

» Knowing the basics of graphing

» Graphing linear and quadratic functions

» Understanding transformations of functions on the *xy*-graph

» Handling higher-order polynomials and circles

Chapter 9

Coordinating a Path through Coordinate Geometry

Coordinate geometry focuses on graphing functions and other equations on the *xy*-graph. Your study of graphing probably began in Algebra I when you plotted points and lines.

Then you continued your graphing training in Algebra II with a more detailed look at linear and quadratic equations.

Lots of ACT questions center on graphing, so in this chapter you get up to speed on this important topic. You begin with a quick review of the basics of the *xy*-graph, and then you move on to the *linear function* — any equation that is graphed as a straight line. I also show you how to use the midpoint formula and the distance formula and how to find the slope of a line. The main event is using the slope-intercept form of the linear function, $y = mx + b$, to answer ACT questions.

Next, you spend some time working with quadratic functions, which take the form $y = ax^2 + bx + c$. I show you some basic qualitative analysis of this function so you can spot its main features easily. I also give you formulas for the axis of symmetry and the vertex. After that, you discover how to work with basic transformation of functions, such as reflections, vertical shifts, and horizontal shifts. To wrap up, I touch on higher-order polynomials and circles.

Reviewing Some Basics of Coordinate Geometry

Coordinate geometry occurs on the *xy-graph* (also called the *coordinate plane* or the *Cartesian plane*), on which each point is assigned a unique pair of coordinates (x, y). Coordinates are identified with a pair of axes — the *x-axis* and the *y-axis*. The two axes are essentially a pair of number lines that cross at (0, 0), a point that's also called the *origin*.

Figure 9-1 shows the basic *xy*-graph with a few points plotted. Each point is identified by a pair of coordinates that includes both an *x*-value and a *y*-value, respectively. When plotting a point, start at the origin, find the *x*-value on the *x*-axis as you would with a number line, and then count either up (positive) or down (negative) to find the *y*-value.

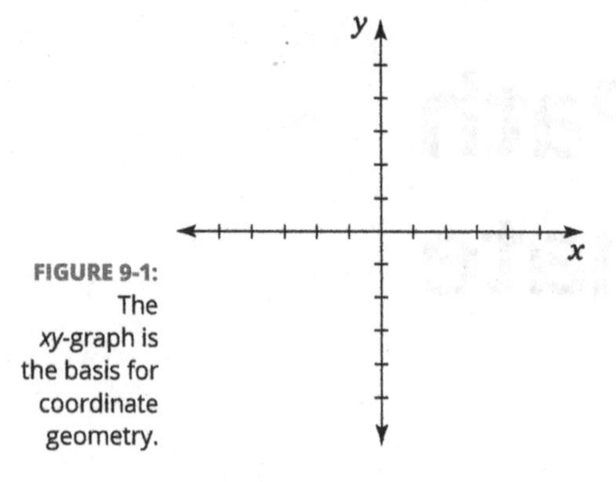

FIGURE 9-1: The *xy*-graph is the basis for coordinate geometry.

Graphing Linear Functions

A *linear function* is simply a line on the *xy*-graph. As in plane geometry (which I discuss in Chapter 11), a line is uniquely defined by two points and extends infinitely in both directions. Graphically, a linear function is a straight line extending infinitely in both directions, as shown in Figure 9-2.

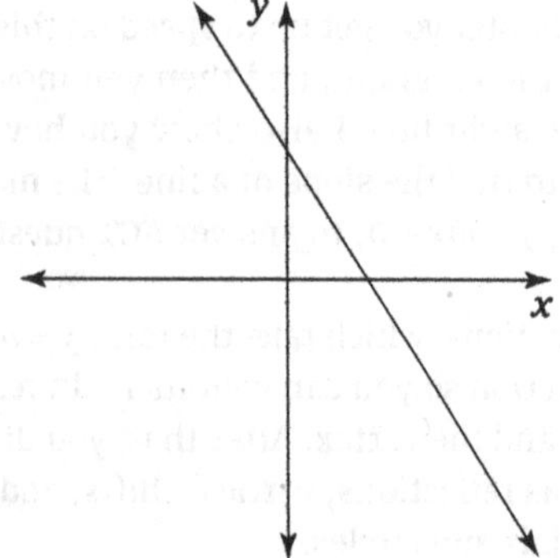

FIGURE 9-2: A linear function is a straight line.

Linear functions are common on the ACT, so understanding them well is important. In this section, I give you the basic skills you need to answer the most common types of ACT questions that involve lines on a graph — whether or not the question actually includes a graph. Flip to Chapter 10 for practice answering questions that use these skills.

Lining up some line segment skills

Two useful formulas for answering ACT questions on coordinate geometry are the *midpoint formula* and the *distance formula.* Both of these formulas provide information about a line segment on a graph between any two points (x_1, y_1) and (x_2, y_2). In this section, I show you how to use both of these formulas.

Finding coordinates with the midpoint formula

The *midpoint formula* allows you to find the coordinates of the midpoint of a line segment between any two points (x_1, y_1) and (x_2, y_2). Here's the formula:

$$\text{Midpoint} = \left(\frac{x_1 + x_2}{2}, \frac{y_1 + y_2}{2}\right)$$

TIP

Here's a little trick to remember the midpoint formula: Notice that for each coordinate it's simply an application of the formula for the mean average of two numbers — the sum of two numbers divided by 2. (For more on finding the mean average, flip to Chapter 6.)

EXAMPLE

In the following graph, N is on $\overline{JK}$, equidistant from J and K. Which of the following are the coordinates of N?

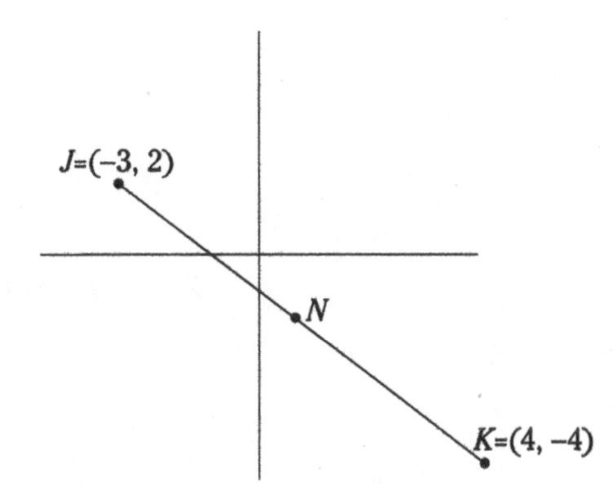

(A) $(1, -1)$

(B) $\left(1, -\frac{3}{2}\right)$

(C) $\left(\frac{1}{2}, -1\right)$

(D) $\left(\frac{1}{2}, -\frac{1}{2}\right)$

(E) $\left(\frac{1}{2}, -\frac{3}{2}\right)$

Plug the four values for the coordinates of J and K into the midpoint formula to get your answer:

$$\text{Midpoint} = \left(\frac{x_1 + x_2}{2}, \frac{y_1 + y_2}{2}\right) = \left(\frac{-3 + 4}{2}, \frac{2 + (-4)}{2}\right) = \left(\frac{1}{2}, -1\right)$$

As you can see, the correct answer is Choice (C).

Measuring length with the distance formula

The *distance formula* allows you to find the length of a line segment between any two points (x_1, y_1) and (x_2, y_2). Consider the formula:

$$\text{Distance} = \sqrt{(x_2 - x_1)^2 + (y_2 - y_1)^2}$$

TIP

The distance formula looks complicated, but it's really just an application of the Pythagorean theorem. To show the similarity between the two, I'll tweak the formula slightly by squaring both sides of it:

$$(\text{Distance})^2 = (x_2 - x_1)^2 + (y_2 - y_1)^2$$

Now notice that $(x_2 - x_1)$ and $(y_2 - y_1)$ are the lengths of the two legs of a triangle with a hypotenuse that's the line segment you want to measure:

$$c^2 = a^2 + b^2$$

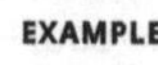

EXAMPLE

In the following figure, what is the length of $\overline{PQ}$?

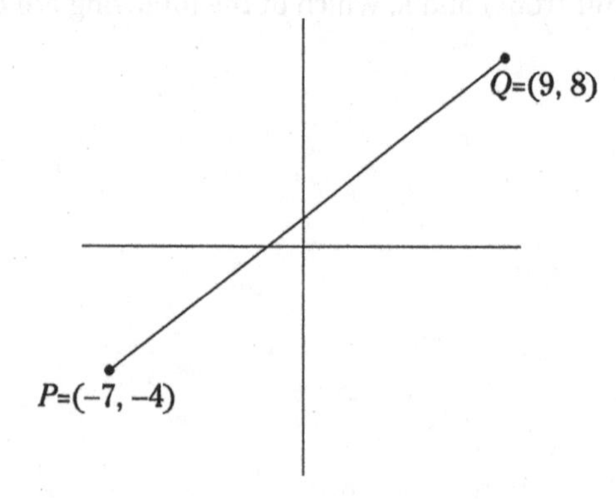

(A) 17

(B) 18

(C) 20

(D) 21

(E) 24

Plug the four values into the distance formula:

$$\text{Distance} = \sqrt{(x_2 - x_1)^2 + (y_2 - y_1)^2} = \sqrt{(9 - (-7))^2 + (8 - (-4))^2}$$

Simplify:

$$= \sqrt{(16)^2 + (12)^2} = \sqrt{256 + 144} = \sqrt{400} = 20$$

Therefore, the correct answer is Choice (C).

Measuring the slope of a line

The *slope* of a line is a measurement of how quickly the line rises or falls as it goes from left to right. When you're working with a graph, the quickest way to measure the slope is with the *rise-run slope formula*:

$$\text{Slope} = \frac{\text{Rise}}{\text{Run}}$$

Find the rise and the run as follows:

- The *rise* is the number of steps up or down from one point to the next.
  - When going up, the rise is a positive number.
  - When going down, the rise is a negative number.
- The *run* is the number of steps from left to right from one point to the next. (The run is *always* a positive number.)

Table 9-1 provides a cheat sheet for evaluating the slope of a line.

TABLE 9-1 Evaluating the Slope of a Line

When the Slope of a Line Is:	The Line (from Left to Right) Is:
A positive number	Rising
A negative number	Falling
0	Horizontal
No slope	Vertical

EXAMPLE

According to the following graph, which of the answer choices has a slope that is an integer?

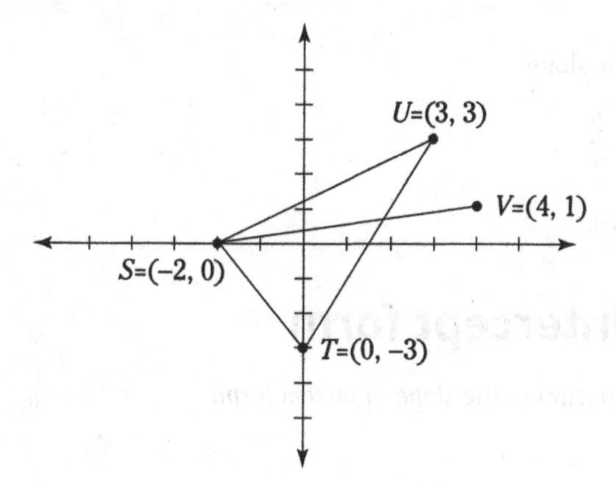

(A) $\overline{ST}$

(B) $\overline{SU}$

(C) $\overline{SV}$

(D) $\overline{TU}$

(E) None of these has a slope that's an integer.

Recall from Chapter 4 that an integer is a whole number that's either positive or negative (or 0). Use the rise-run formula to calculate the slopes of these lines until you find the one whose slope is an integer, or show that none of them is an integer:

$$\text{Slope } \overline{ST} = \frac{\text{Rise}}{\text{Run}} = \frac{-3}{2} = -\frac{3}{2}$$

$$\text{Slope } \overline{SU} = \frac{\text{Rise}}{\text{Run}} = \frac{3}{5}$$

$$\text{Slope } \overline{SV} = \frac{\text{Rise}}{\text{Run}} = \frac{1}{6}$$

$$\text{Slope } \overline{TU} = \frac{\text{Rise}}{\text{Run}} = \frac{6}{3} = 2$$

The correct answer is Choice (D).

Using the rise-run formula can be difficult when you aren't provided a graph of the line you're working with. In these cases, you can use the *two-point slope formula*. This formula allows you to calculate the slope of any line that includes the points (x_1, y_1) and (x_2, y_2). Here's the formula:

$$\text{Slope} = \frac{y_2 - y_1}{x_2 - x_1}$$

WARNING

Notice that in this formula, the *y*-values go on top and the *x*-values on the bottom — don't mix them up!

EXAMPLE

What is the slope of a line that contains the points (–5, 1) and (7, 10)?

(F) $\frac{2}{9}$

(G) $\frac{3}{4}$

(H) $-\frac{2}{9}$

(J) $-\frac{3}{4}$

(K) $-\frac{4}{3}$

Plug these values into the formula for slope:

$$\text{Slope} = \frac{y_2 - y_1}{x_2 - x_1} = \frac{10 - 1}{7 - (-5)} = \frac{9}{12} = \frac{3}{4}$$

Thus, the correct answer is Choice (G).

Applying the slope-intercept form

The most useful form for a linear equation is the *slope-intercept form:*

$$y = mx + b$$

This form contains the two variables *x* and *y*, with *y* isolated on the left side of the equal sign. It also includes two constants represented by the letters *m* and *b*, each of which provides information about the line:

- **The constant *m* is the slope of the line.** For example, in the equation $y = -3x + 5$, the slope of the line is –3.

» **The constant b is the y-intercept of the line.** The *y-intercept* is the point at which the line intersects the y-axis. For example, in the equation $y = -3x + 5$, the y-intercept of the line is 5. In other words, the line includes the point (0, 5).

REMEMBER

You can use the slope-intercept form to help answer virtually every ACT question that involves a line on a graph. If a linear equation isn't given in the slope-intercept form, use algebra to put it in this form.

EXAMPLE

What is the slope of a line whose equation is $3x + 2y = 7$?

(A) 3

(B) $\frac{3}{2}$

(C) $-\frac{3}{2}$

(D) $\frac{7}{2}$

(E) $-\frac{7}{2}$

Begin by putting the equation in the slope-intercept form:

$$3x + 2y = 7$$
$$2y = -3x + 7$$
$$y = -\frac{3}{2}x + \frac{7}{2}$$

In this equation, the slope is $-\frac{3}{2}$, so the correct answer is Choice (C).

You can use the slope-intercept form in conjunction with the formula for slope that I show you in the earlier section "Measuring the slope of a line." The following example shows you how.

EXAMPLE

Which of the following is the equation of a line that has a y-intercept of –4 and goes through the point $(1, -2)$?

(F) $y = x - 4$

(G) $y = 2x - 4$

(H) $y = -2x - 4$

(J) $y = \frac{1}{2}x - 4$

(K) $y = -\frac{1}{2}x - 4$

Remember that the y-intercept is simply a point on the line with an x-coordinate of 0. Because this line has a y-intercept of –4, this point is $(0, -4)$. Plug this point and the point $(1, -2)$ into the formula for slope:

$$\text{Slope} = \frac{y_2 - y_1}{x_2 - x_1} = \frac{-2-(-4)}{1-0} = \frac{2}{1} = 2$$

When you know both the slope $(m = 2)$ and the y-intercept $(b = -4)$, you can write the equation in slope-intercept form:

$$y = 2x - 4$$

So the correct answer is Choice (G).

You also can use the slope-intercept form to answer trickier questions (I told you this form would come in handy). Use the information provided in the problem to find the equation of the line, and then answer the question. The following example walks you through a question like this one.

EXAMPLE

Which of the following is the equation of a line that includes the point $(10, -1)$ and has a slope of $\frac{3}{5}$?

(A) $-x+10y=3$

(B) $10x-y=5$

(C) $3x+5y=9$

(D) $3x-5y=35$

(E) $-3x+5y=7$

To answer this question, you want to use the slope-intercept form, so you need to know the values of m and b. You're given the slope, so $m=\frac{3}{5}$. To find the value of b, plug in the information from the point that's given ($x=10$ and $y=-1$), and then simplify and solve for b:

$$y=mx+b$$
$$-1=\frac{3}{5}(10)+b$$
$$-1=6+b$$
$$-7=b$$

When you know the values of both m and b, you can write the equation for the line in the slope-intercept form:

$$y=\frac{3}{5}x-7$$

Now use algebra to put this equation into the form you need to answer the question:

$$-\frac{3}{5}x+y=-7$$
$$-3x+5y=-35$$
$$3x-5y=35$$

Therefore, the correct answer is Choice (D).

Working with parallel and perpendicular lines

Some ACT questions provide parallel or perpendicular lines on a graph. Remember the following information for when you're dealing with these types of questions:

- *Parallel* lines always have the same slope. For example, any line parallel to $y=-\frac{2}{5}x+8$ also has a slope of $-\frac{2}{5}$.
- *Perpendicular* lines always have slopes that are *negative reciprocals* of each other. To find the negative reciprocal, turn the fraction upside-down and negate it — that is, if it's positive, change it to negative, and if it's negative, change it to positive. For example, any line perpendicular to $y=-\frac{2}{5}x+8$ has a slope of $\frac{5}{2}$.

What is the equation of a line that has an x-intercept of -3 and is parallel to $y = 3x - 9$?

EXAMPLE

(A) $y = 3x + 3$

(B) $y = 3x - 3$

(C) $y = 3x + 9$

(D) $y = -3x - 3$

(E) $y = -3x - 9$

The line you're looking for is parallel to $y = 3x - 9$, so it also has a slope of 3. Thus, you can rule out Choices (D) and (E). It has an x-intercept of -3, which is the point $(-3, 0)$. Plug these values into the slope-intercept form and solve for b, as I show you in the earlier section "Applying the slope-intercept form." Here's what your work should look like:

$$y = mx + b$$
$$0 = 3(-3) + b$$
$$0 = -9 + b$$
$$9 = b$$

Now that you know both the slope and the y-intercept for this line, you can write the equation in the slope-intercept form:

$$y = mx + b$$
$$y = 3x + 9$$

The correct answer is Choice (C).

What is the equation of a line that passes through the point (0, 3) and is perpendicular to $y = 8x + 5$?

EXAMPLE

(F) $y = 8x$

(G) $y = 8x + 3$

(H) $y = -\frac{1}{8}x$

(J) $y = -\frac{1}{8}x + 3$

(K) $y = -\frac{1}{8}x + 5$

This line you're looking for is perpendicular to $y = 8x + 5$, so its slope is the negative reciprocal of 8. In other words, $m = -\frac{1}{8}$. Thus, you can rule out Choices (F) and (G). You could plug this value, along with $x = 0$ and $y = 3$, into the slope-intercept form to find b, but you can use an even easier way. Notice that (0, 3) is the y-intercept, so $b = 3$. Plug in these values directly:

$$y = mx + b$$
$$y = -\frac{1}{8}x + 3$$

So the correct answer is Choice (J).

Giving Your Quads a Workout: Graphing Quadratic Functions

A *quadratic function* is any function in the following form:

$$y = ax^2 + bx + c$$

When you graph a quadratic function on the *xy*-plane, the result is a *parabola* — a curved shape extending infinitely in both directions, as shown in Figure 9-3.

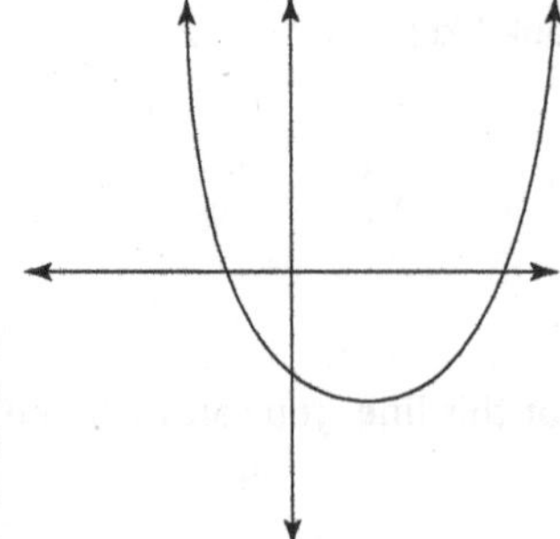

FIGURE 9-3: A quadratic function is graphed as a parabola.

The quadratic equation, $ax^2 + bx + c = 0$, which you encounter in Chapter 8, is a form of the quadratic function in which $y = 0$. In this section, I show you everything you need to know to successfully graph quadratic functions.

TIP

If you have a graphing calculator, such as the TI-83 or TI-89, you can use it to graph quadratic functions. For more details, pick up *TI-83 Plus Graphing Calculator For Dummies* by C.C. Edwards (Wiley) or *TI-89 Graphing Calculator For Dummies* by C.C. Edwards (Wiley).

Searching for the qualities of a quadratic

Sketching a quadratic equation by plotting points can be time-consuming. If you have a graphing calculator, you can use it to graph a quadratic function. However, you can answer some ACT questions even more easily if you know a few simple tricks. Each of these tricks links the coefficients (*a*, *b*, and *c*) in a quadratic equation to a different *quality* of the parabola — that is, a general rule about how the parabola is placed on the graph. I show you the three tricks you need to know in the following sections.

Trick 1: Checking the direction of your parabola

This first trick relates to the sign of the variable *a* (in the term ax^2):

- When *a* is positive, the graph is *concave up*. In other words, you can imagine pouring water in it like a cup.
- When *a* is negative, the graph is *concave down*. In this case, the cup is upside-down.

This trick is especially helpful when a question gives you the graph of a parabola, because it's easy to see at a glance which direction it's facing.

EXAMPLE

Which of the following equations CANNOT be the graph of the function shown here?

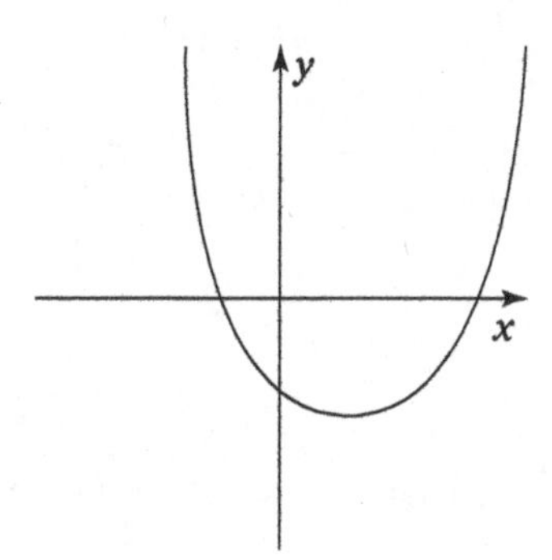

(A) $y=-x^2-x-1$

(B) $y=100x^2-1000x-100$

(C) $y=0.01x^2-0.1x-0.001$

(D) $y=\frac{3}{5}x^2-6x-7$

(E) $y=x^2-\frac{\pi}{2}x-\pi$

Even with a graphing calculator, graphing all five of these equations would take a long time. Fortunately, the trick I show you earlier provides a much easier way to answer the question: The parabola in the figure is concave up, so a is positive. Voila! So the equation for this graph isn't $y=-x^2-x-1$, making the correct answer Choice (A).

Trick 2: Looking at your parabola's vertex

The second trick relates to the signs of the variables a and b (in the terms ax^2 and bx):

- **When the signs of *a* and *b* are the same (either both positive or both negative), the graph *shifts to the left.*** That is, the vertex of the parabola is to the left of the *y*-axis.
- **When the signs of *a* and *b* are different (one is positive and the other is negative), the graph *shifts to the right.*** In this case, the vertex of the parabola is to the right of the *y*-axis.

TIP

A good mnemonic for remembering this rule is that the word *same* and the word *left* both have four letters.

EXAMPLE

A quadratic function $y=ax^2+bx+c$ crosses the x-axis at $x=4$ and $x=-2$. Which of the following must be true?

(A) $a>0$

(B) $a \geq b$

(C) $a \neq b$

(D) $b \neq c$

(E) $c<0$

At first glance, you may think this difficult question isn't even answerable. A quick sketch shows you that a lot of different parabolas could fit this description:

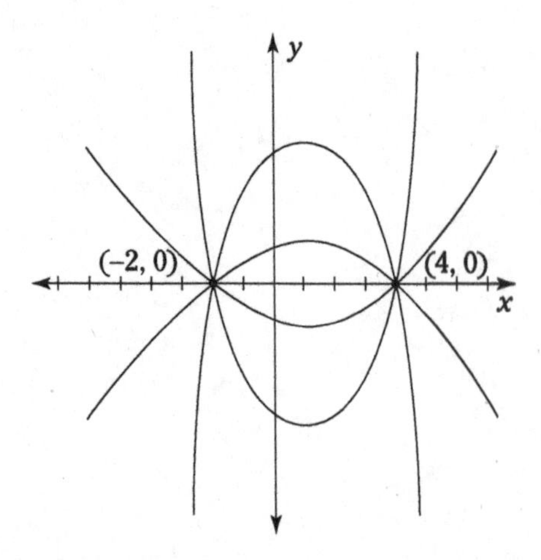

The only thing you know for sure is that the parabola's vertex is horizontally in the middle of these two points, so it's somewhere on the line $x = 1$. Thus, the parabola (whatever it looks like) is shifted to the right, so a and b have different signs. So $a \neq b$, making the correct answer Choice (C).

Trick 3: Keeping an eye on variable c

The third trick relates to the sign of the variable c (the constant term of the equation). Keep the following rules in mind:

- **When c is positive, the y-intercept is positive.** In other words, the parabola intersects the y-axis above the origin.
- **When c is negative, the y-intercept is negative.** That is, the parabola intersects the y-axis below the origin.

WARNING

Be clear that in a quadratic function, c is the y-intercept. In contrast, in a linear function ($y = mx + b$), b is the y-intercept.

EXAMPLE

Which of the following could be a graph of the function $y = -x^2 + 5x - 2$?

(A)

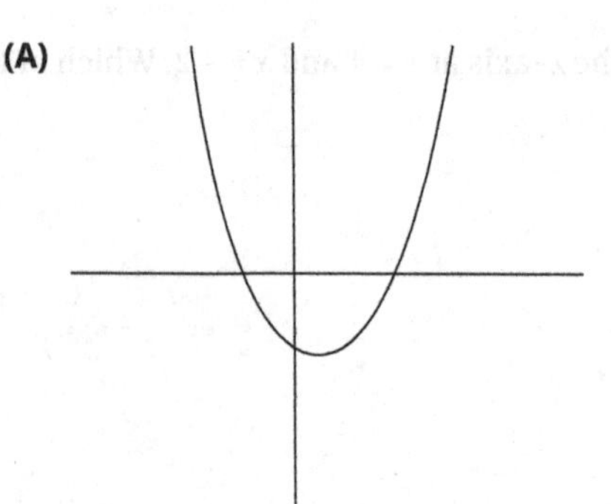

(B)

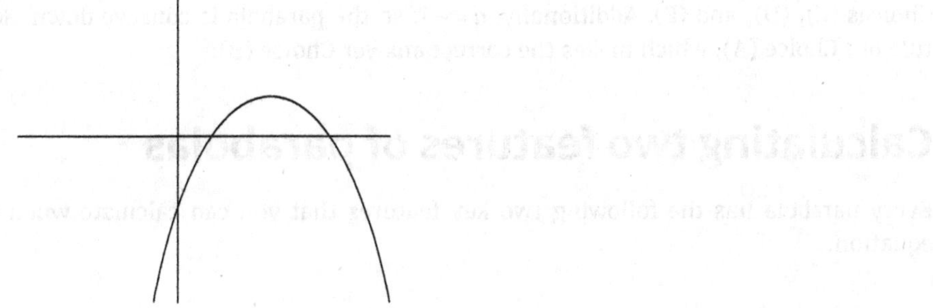

(C)

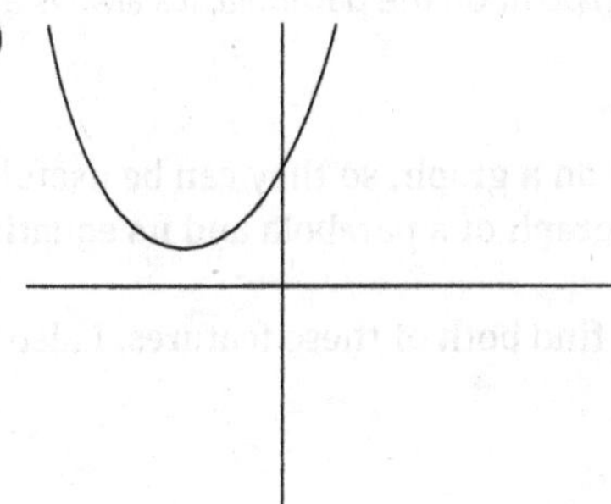

(D)

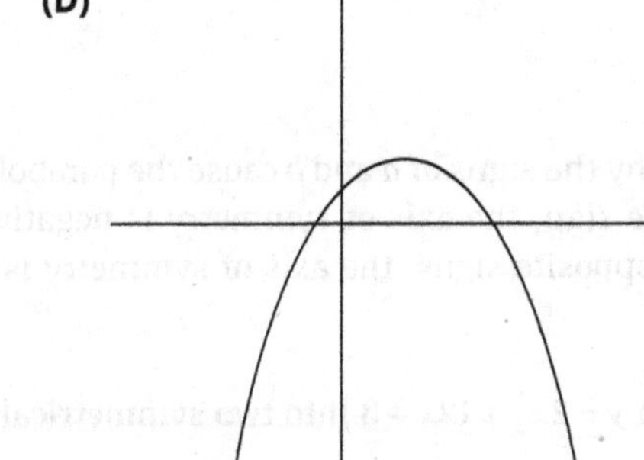

(E)

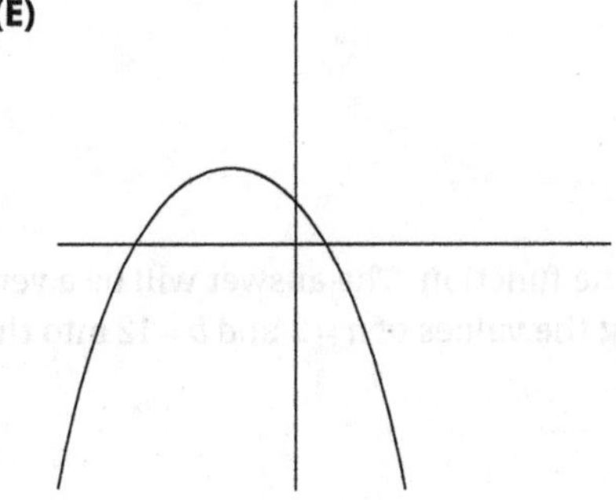

In this equation, $c = -2$, so the y-intercept is below the y-axis. As a result, you can rule out Choices (C), (D), and (E). Additionally, $a = -1$, so the parabola is concave down. So you also can rule out Choice (A), which makes the correct answer Choice (B).

Calculating two features of parabolas

Every parabola has the following two key features that you can calculate when you know the equation:

- **The axis of symmetry:** This axis is the vertical line that divides the parabola down the middle.
- **The vertex:** The vertex is the lowest (or highest) point on the parabola. It's always a point on the axis of symmetry.

These two features of a parabola are easy to spot on a graph, so they can be useful when a question asks you to make a connection between the graph of a parabola and its equation.

In this section, I provide the formulas you use to find both of these features. I also show you how to apply them to answer ACT questions.

Accessing the axis of symmetry

The *axis of symmetry* is the vertical line that divides a parabola down the middle. For a quadratic function $y = ax^2 + bx + c$, the formula for the axis of symmetry is

$$x = -\frac{b}{2a}$$

REMEMBER

The formula for the axis of symmetry explains why the signs of a and b cause the parabola to shift to the left or right. When a and b have the same sign, the axis of symmetry is negative, which shifts the parabola to the left; when they have opposite signs, the axis of symmetry is positive, shifting the parabola to the right.

EXAMPLE

Which of the following lines divides the function $y = 2x^2 + 12x + 3$ into two symmetrical parts?

(A) $x = 2$

(B) $x = -3$

(C) $x = 12$

(D) $y = -2$

(E) $y = 3$

The question asks for the axis of symmetry for the function. The answer will be a vertical line, so you can rule out Choices (D) and (E). Simply plug the values of $a = 2$ and $b = 12$ into the formula to calculate the axis of symmetry:

$$x = -\frac{b}{2a}$$
$$x = -\frac{12}{2(2)}$$
$$x = -3$$

Therefore, the correct answer is Choice (B).

Changing direction at the vertex

The *vertex* of a parabola is the point where the parabola changes directions. It's always either the lowest or the highest point on the parabola. Here's the formula for the coordinates of the vertex for the quadratic function $y = ax^2 + bx + c$:

$$\text{Vertex} = \left(-\frac{b}{2a}, -\frac{b^2 - 4ac}{4a}\right)$$

TIP

This formula can be useful but difficult to remember. To help yourself remember, notice that the x-coordinate is simply the x-value of the axis of symmetry, which I discuss in the earlier section, "Accessing the axis of symmetry." The y-coordinate looks complicated, but you may recall the numerator $b^2 - 4ac$ as the discriminant, which is the part of the quadratic equation (see Chapter 8) that's inside the square root.

EXAMPLE

In the following figure, the vertex of the parabola $y = -2x^2 + 3x + 1$ is located at (h, k). What is the value of k?

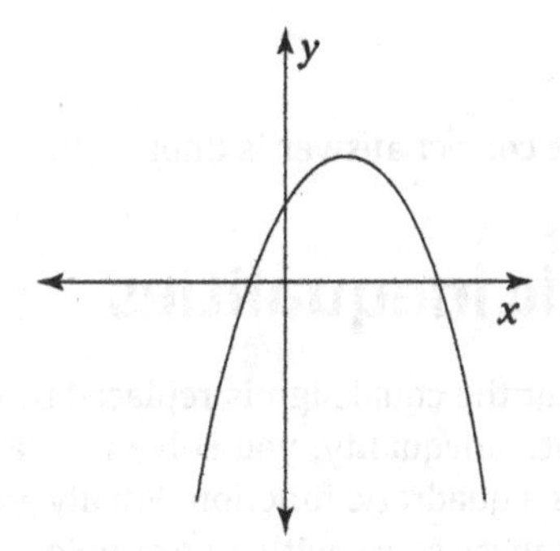

(A) $\frac{17}{4}$

(B) $-\frac{17}{4}$

(C) $\frac{17}{8}$

(D) $-\frac{17}{8}$

(E) $\frac{25}{8}$

The question asks you for the value of k, which is the y-coordinate of the vertex. Plug $a = -2$, $b = 3$, and $c = 1$ into the formula to find the value:

$$k = -\frac{b^2 - 4ac}{4a} = -\frac{3^2 - 4(-2)(1)}{4(-2)} = -\frac{9 + 8}{-8} = \frac{17}{8}$$

As you can see, the correct answer is Choice (C).

As an alternative strategy to the formula that I gave you above, another way to find the vertex of a parabola is to plug in the value of the axis of symmetry for x and solve for y. For example:

EXAMPLE

On the xy-plane, what is the vertex of the parabola whose equation is $y = x^2 + 10x + 17$?

(F) (5, 42)

(G) (5, 92)

(H) (−5, −42)

(J) (−5, −8)

(K) (−5, 42)

First, find the axis of symmetry:

$$x = -\frac{10}{2(1)} = -5$$

Plug this value into the equation and solve for y:

$$y = (-5)^2 + 10(-5) + 17 = 25 - 50 + 17 = -8$$

Thus, the coordinates of the vertex are $(-5, -8)$, so the correct answer is Choice (J).

Solving and graphing quadratic inequalities

A *quadratic inequality* resembles a quadratic equation, but the equal sign is replaced by one of the four inequality signs (<, >, ≤, or ≥). To solve a quadratic inequality, you solve it as a quadratic equation, as I discuss in Chapter 8, and then graph it as a quadratic function. Finally, you use the graph to answer the question. This process makes a lot more sense with an example.

EXAMPLE

If $x^2 - 7x + 12 < 0$, then one possible value of x is

(A) $\frac{8}{13}$

(B) $\frac{10}{3}$

(C) $\frac{14}{3}$

(D) $\frac{17}{3}$

(E) $\frac{20}{3}$

Begin by solving the related quadratic equation:

$$x^2 - 7x + 12 = 0$$
$$(x - 3)(x - 4) = 0$$

Now split into two equations and solve both:

$$x - 3 = 0 \qquad x - 4 = 0$$
$$x = 3 \qquad x = 4$$

At this point, a bit of graphing helps. The two values of x (in this case, 3 and 4) also are called *zeros* of the equation, because the graph of the function $y = x^2 - 7x + 12$ crosses the x-axis at these points. Additionally, the parabola is concave up because the x^2 term is positive. These two facts enable you to sketch a graph that's useful for answering the question:

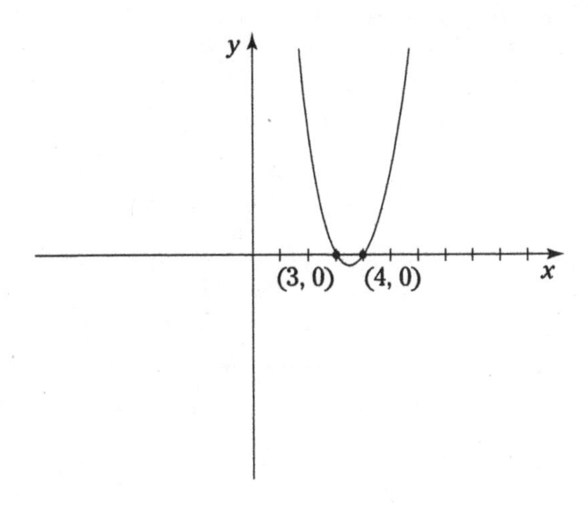

The inequality $x^2 - 7x + 12 < 0$ holds true wherever the graph is *below* the x-axis — that is, when $3 < x < 4$. Therefore, the inequality is true when $x = \frac{10}{3}$. So the correct answer is Choice (B).

Recognizing Three Transformations

A *transformation* is a small change to a function that alters the graph of that function in a predictable way. The transformations you're likely to see on the ACT come in three varieties: reflections, vertical shifts, and horizontal shifts. In this section, I show you how to work with all three of these transformations.

Reversing graphs with reflections

A *reflection* of a function causes the graph of that function to appear reversed on the opposite side of either the x-axis or the y-axis. Here are the two types of reflections for the function $y = f(x)$:

- **The function $y = -f(x)$ reflects the function $y = f(x)$ vertically across the x-axis, negating the whole function.**
- **The function $y = f(-x)$ reflects the function $y = -f(x)$ horizontally across the y-axis, changing each x to $-x$.**

EXAMPLE

In the following figure, $f(x) = x^2 + 3$ and $g(x)$ is a reflection of this function across the x-axis, so $g(x) =$

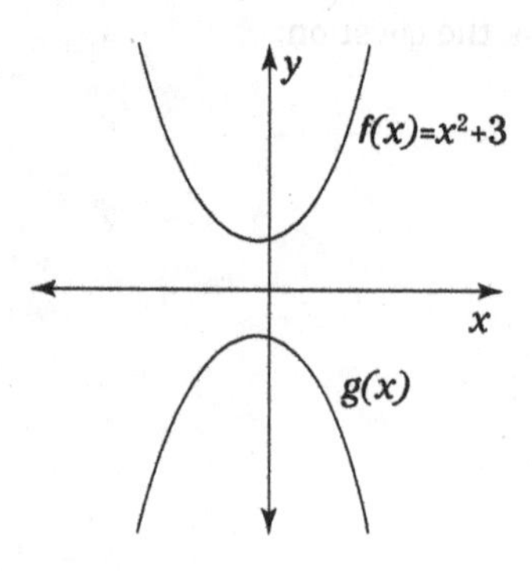

(A) $x^2 - 3$
(B) $-x^2 + 3$
(C) $-x^2 - 3$
(D) $(-x)^2 + 3$
(E) $(-x)^2 - 3$

The function $g(x)$ is a reflection of $f(x)$ vertically across the x-axis. Use parentheses so you don't get confused, and then simplify:

$$g(x) = -f(x) = -(x^2 + 3) = -x^2 - 3$$

As you can see, the correct answer is Choice (C).

EXAMPLE

Which of the following would be a reflection of the function $y = x^3 - x^2$ horizontally across the y-axis?

(F) $y = x^2(x + 1)$
(G) $y = x^2(1 - x)$
(H) $y = -x^2(x + 1)$
(J) $y = -x^2(x - 1)$
(K) $y = -x^2(1 - x)$

To reflect $y = x^3 - x^2$ horizontally across the y-axis, change every x to $-x$ (use parentheses to keep track of these changes) and then simplify:

$$y = (-x)^3 - (-x)^2 = -x^3 - x^2$$

To make this result resemble the answers provided, factor out $-x^2$:

$$y = -x^2(x + 1)$$

Therefore, the correct answer is Choice (H).

Movin' on up (or down): Vertical shifts

A *vertical shift* moves a function either up or down. Here are the two types of vertical shifts for the function $y = f(x)$:

- **The function $y = f(x) + n$ vertically shifts the function up n units.** Here you add n to the whole function.
- **The function $y = f(x) - n$ vertically shifts the function down n units.** In other words, to shift down, you subtract n from the whole function.

EXAMPLE

If $f(x) = x^2 + 2x + 4$, and $g(x)$ shifts this function up 1 unit, which of the following equals $g(x)$?

(A) $x^2 + 4$

(B) $x^2 + 2x + 3$

(C) $x^2 + 2x + 5$

(D) $x^2 + 4x + 5$

(E) $x^2 + 4x + 6$

To shift the function up 1 unit, just add 1 to the entire function:

$$g(x) = (x^2 + 2x + 4) + 1 = x^2 + 2x + 5$$

So the correct answer is Choice (C).

Shifting horizontally

A *horizontal shift* moves a function either to the left or to the right. Here are the two types of horizontal shifts for the function $y = f(x)$:

- **The function $y = f(x + n)$ horizontally shifts the function to the left, or in the negative direction, by n units.** You change each x to $x + n$.
- **The function $y = f(x - n)$ horizontally shifts the function to the right, or in the positive direction, by n units.** You change each x to $x - n$.

WARNING

Notice that horizontal shift occurs in the opposite direction than you may think: Adding inside the parentheses moves the function in the negative direction along the x-axis.

EXAMPLE

If $f(x) = 3x^2 + 5x - 1$, and $g(x)$ shifts this function 2 units to the left, then $g(x) =$

(A) $3x^2 + 5x + 1$

(B) $3x^2 + 5x - 3$

(C) $3x^2 - 7x + 1$

(D) $3x^2 + 17x + 1$

(E) $3x^2 + 17x + 21$

To shift $f(x) = 3x^2 + 5x - 1$ to the left 2 units (that is, in the negative direction), change every x to $x + 2$:

$$g(x) = 3(x + 2)^2 + 5(x + 2) - 1$$

Now simplify to get your answer:

$$= 3(x^2 + 4x + 4) + 5x + 10 - 1$$
$$= 3x^2 + 12x + 12 + 5x + 10 - 1$$
$$= 3x^2 + 17x + 21$$

Thus, the correct answer is Choice (E).

EXAMPLE

On the xy-graph, $g(x)$ shifts the function $f(x)$ up 5 units and to the right 4 units. Which of the following is true?

(F) $g(x) = f(x + 4) + 5$
(G) $g(x) = f(x + 5) + 4$
(H) $g(x) = f(x + 4) - 5$
(J) $g(x) = f(x - 4) + 5$
(K) $g(x) = f(x - 5) - 4$

To shift the function 5 units up, add 5 to the entire function (as I show you in the earlier section "Movin' on up (or down): Vertical shifts"):

$$f(x) + 5$$

To shift the function 4 units to the right (that is, in the positive direction), change x to $x - 4$:

$$f(x - 4) + 5$$

The correct answer is Choice (J).

Taking on Advanced Equations

Most ACT questions that cover coordinate geometry focus on linear and quadratic equations, which I discuss earlier in this chapter. In this section, however, I touch on two advanced types of equations: *polynomials* and *circles*.

Raising the bar with higher-order polynomials

Linear functions and quadratic functions are specific examples of *polynomials* — functions based on raising x to the power of a positive integer. The *degree* of a polynomial is the highest exponent to which x is raised. Table 9-2 shows some common polynomials and the degrees to which they are raised.

TABLE 9-2: Common Polynomials

Degree	Name	Example
1st degree	Linear equation	$y = 4x + 2$
2nd degree	Quadratic equation	$y = x^2 + 5x - 6$
3rd degree	Cubic equation	$y = x^3 - 3x^2 - 1$
4th degree	Quartic equation	$y = 16x^4 - 1$

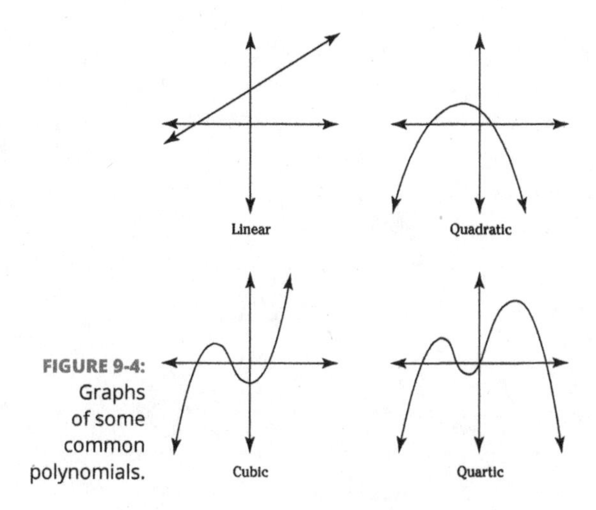

FIGURE 9-4: Graphs of some common polynomials.

Higher-order polynomials aren't so common on the ACT, but knowing a few basics is a good idea. Figure 9-4 shows you example graphs of basic first- through fourth-degree polynomials. The degree of a polynomial tells you the maximum number of zeros for that equation, or, in other words, the maximum number of times it can cross the x-axis.

EXAMPLE

Which of the following functions or combinations of functions CANNOT intersect with the x-axis exactly four times?

(A) Four linear functions

(B) A linear function and a quadratic function

(C) Two quadratic functions

(D) A fourth-degree polynomial

(E) A fifth-degree polynomial

Each of the preceding functions or combination of functions can cross the x-axis exactly four times *except* the combination of a linear function and a quadratic function, which can cross the x-axis no more than three times. Therefore, the correct answer is Choice (B).

EXAMPLE

Which of the following CANNOT be the graph of a polynomial?

(F)

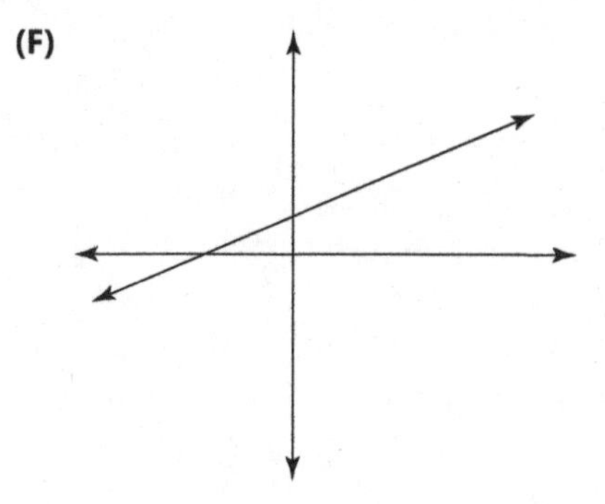

(G)

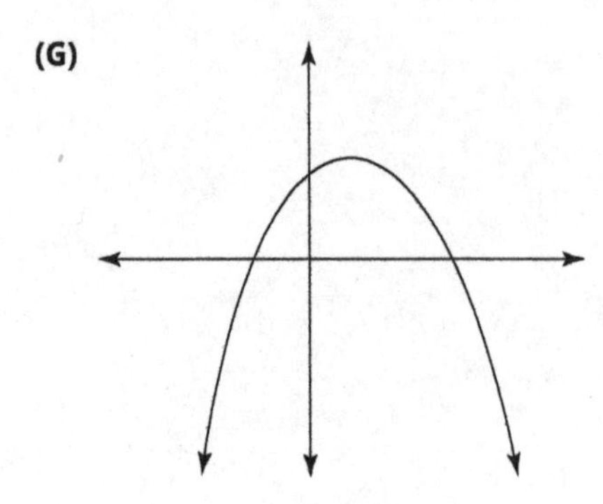

(H)

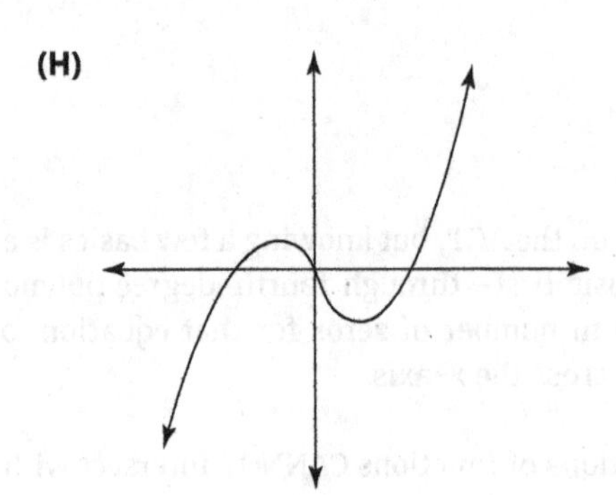

(J)

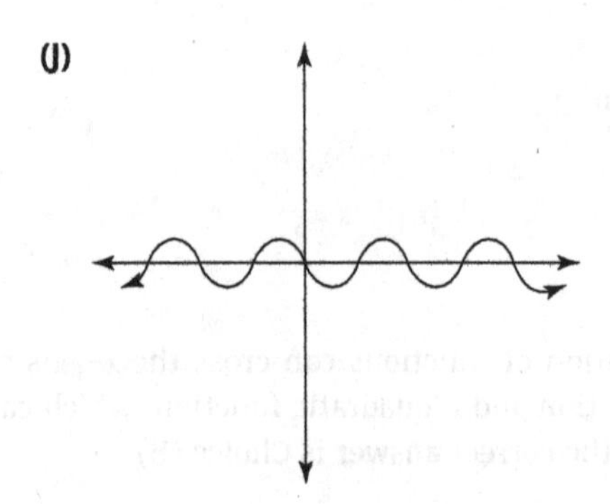

(K)

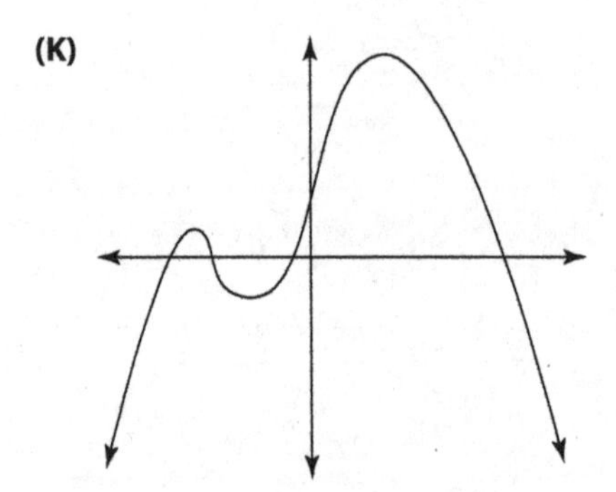

The sine curve crosses the x-axis an infinite number of times, so it can't be a polynomial of any degree. Thus, the correct answer is Choice (J).

Going 'round and 'round with circles

Although the circle is a familiar and simple shape, it's less commonly associated with coordinate geometry. ACT questions involving circles are rather uncommon, but they do show up from time to time. Most ACT questions focusing on circles involve a straightforward application of the basic formula I introduce in this section.

The formula for a circle of radius r centered at the point (h, k) is:

$$(x-h)^2+(y-k)^2=r^2$$

For example, here's the equation for a circle with a radius of 5 centered at $(3,-4)$:

$$(x-3)^2+(y+4)^2=25$$

WARNING

Be careful not to mix up positive and negative coordinates in the formula for a circle: In this example, $(x-3)^2$ moves the circle along the x-axis in the *positive* direction, and $(y+4)^2$ moves it along the y-axis in the *negative* direction.

EXAMPLE

What is the formula for the circle shown in the following figure given that the circle has a radius of 4?

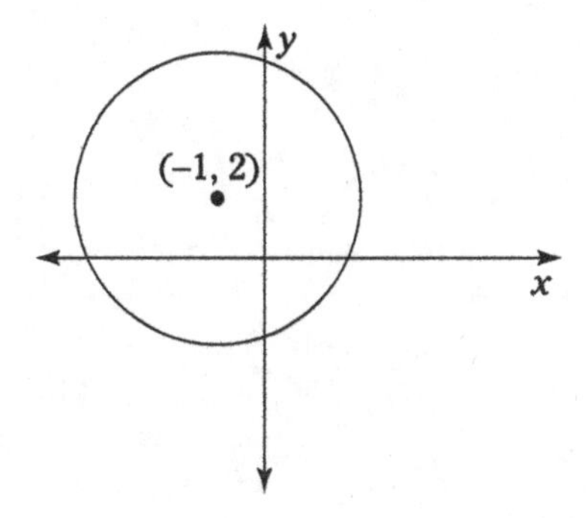

(A) $(x+1)^2+(y-2)^2=4$
(B) $(x-1)^2+(y+2)^2=4$
(C) $(x+1)^2+(y-2)^2=16$
(D) $(x-1)^2+(y+2)^2=16$
(E) $(x-1)^2+(y-4)^2=16$

Place the values $h=-1$, $k=2$, and $r=4$ into the formula for a circle and simplify:

$$(x-h)^2+(y-k)^2=r^2$$
$$(x-\ (-1))^2+(y-2)^2=4^2$$
$$(x+1)^2+(y-2)^2=16$$

So the correct answer is Choice (C).

Chapter 10
Practice Problems for Intermediate Algebra and Coordinate Geometry

Ready for some practice questions on intermediate algebra and coordinate geometry from the material in Chapters 8 and 9? Sure you are! This chapter provides you with plenty of practice as well as detailed explanations of how to arrive at the correct answers. If you find you're struggling with any particular concepts, refer to Chapter 8 or 9 again for some more review.

Intermediate Algebra and Coordinate Geometry Practice Problems

Here are 36 questions that cover what you need to know about inequalities, systems of equations, graphing, functions, and much more. When you're finished with the questions, check your answers later in the chapter.

1. Which of the following is equivalent to the inequality $-5(3x-2)<25-12x$?

(A) $x>5$

(B) $x<5$

(C) $x>-5$

(D) $x<-5$

(E) $x\leq 5$

2. In the following figure, which line segment has a slope of $-\frac{1}{3}$?

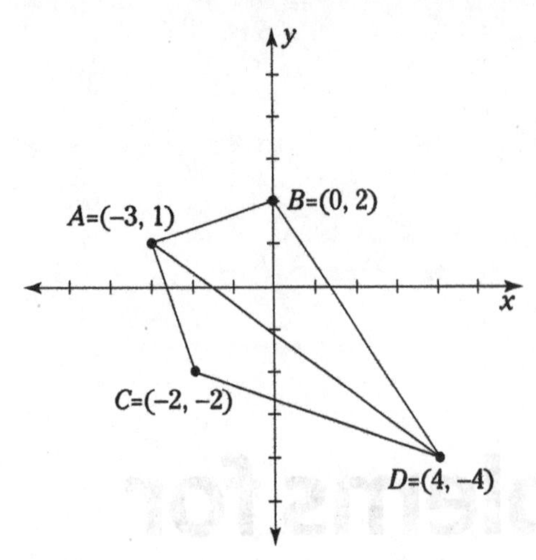

(F) $\overline{AB}$

(G) $\overline{AC}$

(H) $\overline{AD}$

(J) $\overline{BD}$

(K) $\overline{CD}$

3. Anika and Bette are both celebrating their birthdays today. If Anika were ten years older and Bette were three years younger, then Anika would be twice as old as Bette. And if Anika were three years older and Bette were twice her current age, Anika's age would be $\frac{3}{4}$ that of Bette. How old is Bette?

(A) 26

(B) 28

(C) 32

(D) 35

(E) 36

4. If a and b are the two values that satisfy the equation $x^2 - 9x + 20 = 0$, what is the value of $a^2 + b^2$?

(F) 25

(G) 41

(H) 100

(J) 104

(K) 401

5. What is the formula of a line that is perpendicular to $y = 5x + 11$ and includes the point (0, 9)?

(A) $y = 5x + 9$

(B) $y = \frac{1}{5}x$

(C) $y = \frac{1}{5}x + 11$

(D) $y = -\frac{1}{5}x + 9$

(E) $y = -\frac{1}{5}x + 11$

6. If $f(x) = 8x^2$ and $g(x) = 2x - 3$, what is the value of $f\left(\frac{1}{2}\right) - g(10)$?

(F) -1

(G) -9

(H) -15

(J) -19

(K) -21

7. In the following figure, what would be the midpoint of $\overline{MN}$?

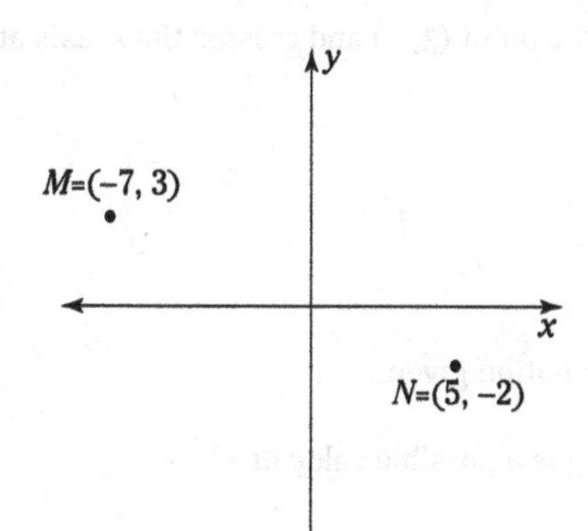

(A) $\left(-1, \frac{1}{3}\right)$

(B) $\left(-1, \frac{1}{2}\right)$

(C) $\left(-1, \frac{3}{2}\right)$

(D) $\left(\frac{1}{2}, -1\right)$

(E) $\left(\frac{1}{3}, -1\right)$

8. In the following figure, if $g(x)$ is the reflection across the x-axis of $f(x)=2x^2+1$, then $g(x)=$

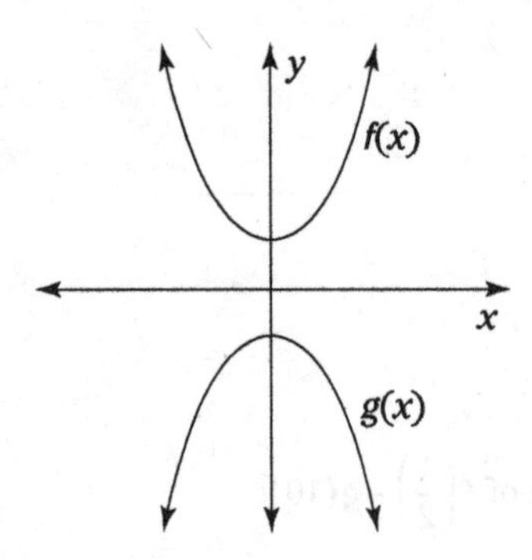

(F) $2x^2+1$

(G) $2x^2-1$

(H) $-2x^2+1$

(J) $-2x^2-1$

(K) $2(-x)^2+1$

9. What is the slope of a line that includes the point (3, 2) and crosses the x-axis at $x=9$?

(A) 3

(B) –3

(C) $\frac{1}{3}$

(D) $-\frac{1}{3}$

(E) Cannot be determined from the information given.

10. If $3x^2+12x+4=0$, which of the following is a possible value of x?

(F) $-2+\frac{1}{4}\sqrt{3}$

(G) $-2+\frac{1}{3}\sqrt{3}$

(H) $-2+\frac{2}{3}\sqrt{6}$

(J) $-2+\frac{1}{3}\sqrt{30}$

(K) $-2+\frac{1}{6}\sqrt{30}$

11. What is the formula of a line with a slope of –2 that includes the point (7, 1)?

(A) $y=-2x-13$

(B) $y=-2x+15$

(C) $y=-2x+\frac{2}{7}$

(D) $y=-2x-\frac{7}{2}$

(E) $y=-2x-\frac{1}{14}$

12. Two values a and b are inversely proportional such that if $a=3$, then $b=10$. Which of the following is true?

(F) If $a=1$, then $b=\frac{10}{3}$

(G) If $a=5$, then $b=6$

(H) If $a=6$, then $b=20$

(J) If $a=6.5$, then $b=6.5$

(K) If $a=30$, then $b=0$

13. In the following figure, what is the length of $\overline{UV}$?

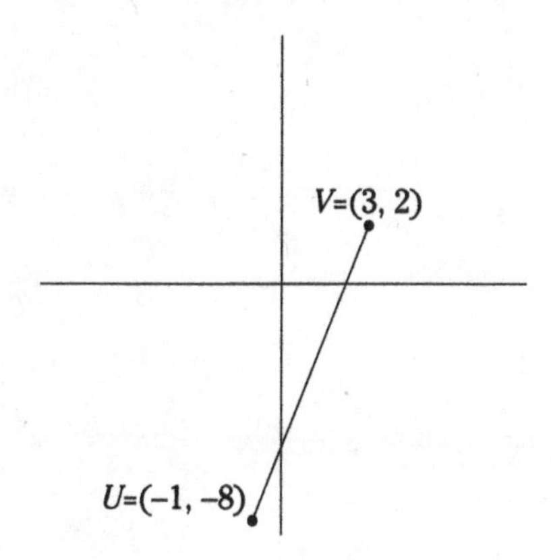

(A) $2\sqrt{13}$

(B) $4\sqrt{13}$

(C) $2\sqrt{26}$

(D) $2\sqrt{29}$

(E) $4\sqrt{29}$

14. Which of the following could be the graph of the equation $y = 3x^2 - 10x - 2$?

(F)

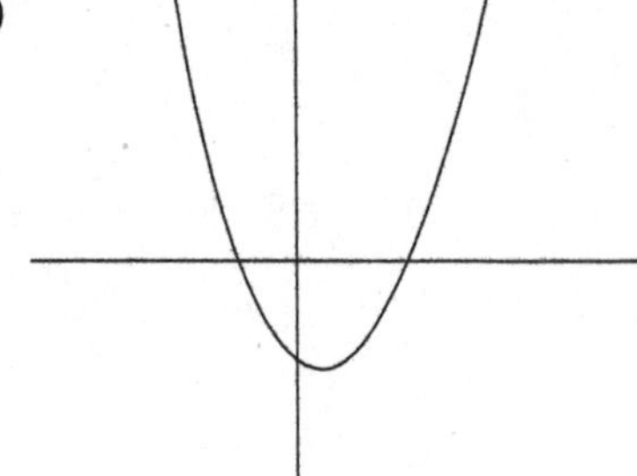

(G)

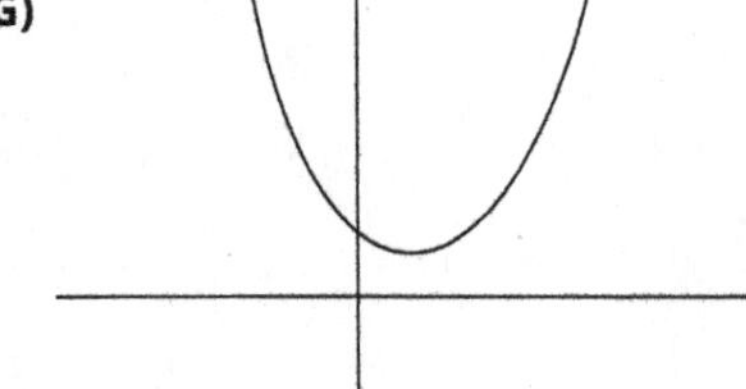

(H)

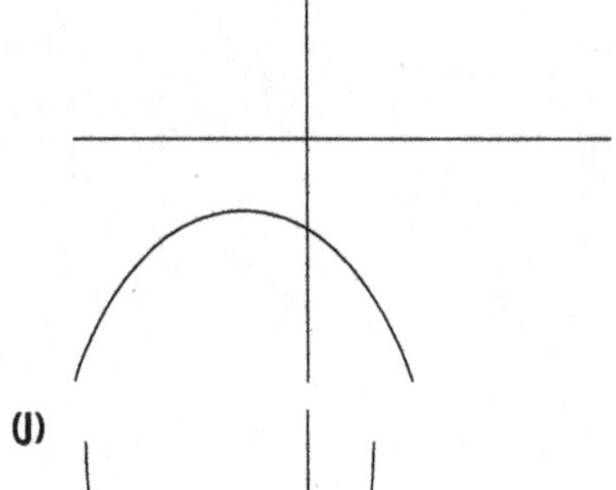

(J)

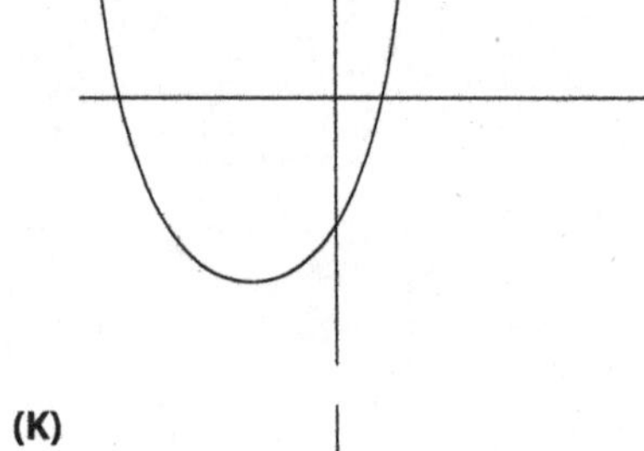

(K)

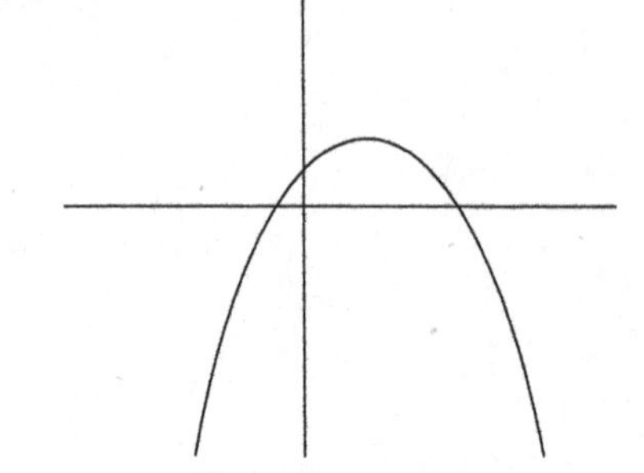

15. Which of the following is the solution set for n for the inequality $|5-2n|>9$?

(A) $2<n<7$

(B) $-2<n<7$

(C) $-7<n<2$

(D) $n<-2$ or $n>7$

(E) $n<-7$ or $n>2$

16. Which of the following transformations shifts $y=x^2+2x+3$ horizontally one unit to the right?

(F) $y=x^2+2$

(G) $y=x^2+3$

(H) $y=x^2+2x+2$

(J) $y=x^2+2x+4$

(K) $y=x^2+4x+6$

Questions 17 and 18 are based on the following information: The Kleen-Pro Housecleaning Company charges a flat rate of $150 for an initial visit of four hours and $20 per hour for each subsequent visit.

17. Which of the following functions outputs the dollar cost of x hours of service by Kleen-Pro for any value of $x \geq 4$?

(A) $f(x)=20x+70$

(B) $f(x)=20x-70$

(C) $f(x)=20x+150$

(D) $f(x)=20x-150$

(E) $f(x)=20x+220$

18. What is the cost of 20 hours of Kleen-Pro service?

(F) $250

(G) $300

(H) $400

(J) $470

(K) $550

19. If $g(x)$ shifts the function $f(x)$ three units down and six units to the left, then $g(x)=$

(A) $f(x+3)-6$

(B) $f(x-3)+6$

(C) $f(x-3)-6$

(D) $f(x+6)-3$

(E) $f(x-6)-3$

20. Which of the following are the coordinates of the vertex for the parabola whose equation is $y = x^2 + 6x + 5$?

(F) $(3, 4)$

(G) $(-3, 4)$

(H) $(-3, -4)$

(J) $(4, -3)$

(K) $(-4, -3)$

21. Which of the following values of x does NOT satisfy the inequality $2x^2 - x - 10 < 0$?

(A) $-\frac{12}{5}$

(B) $-\frac{7}{5}$

(C) $-\frac{3}{5}$

(D) $\frac{4}{5}$

(E) $\frac{11}{5}$

22. On an xy-graph, line r and line s are perpendicular to each other, line r intersects $(0, 11)$, and the equation of line s is $y = 3x + 1$. At what point do line r and line s intersect?

(F) $(3, 8)$

(G) $(3, 9)$

(H) $(3, 10)$

(J) $(4, 9)$

(K) $(4, 10)$

23. If $3p + 4q = -14$ and $7p + 5q = 15$, what is the value of $p + q$?

(A) -1

(B) -2

(C) -3

(D) -4

(E) -5

24. If Equation 1 is $y = x + 1$ and Equation 2 is $y = x^3$, which of the following is true?

(F) Equation 1 is quadratic.

(G) Equation 1 is cubic.

(H) Equation 2 is linear.

(J) Equation 2 is quadratic.

(K) Equation 2 is cubic.

25. Which of the following values is in the domain of the function $f(x) = \frac{\sqrt{8 - x^3}}{1 - x}$?

(A) 1

(B) 2

(C) 3

(D) 4

(E) 5

26. In the following figure, the equation for the parabola is $y = x^2 + 1$. The line crosses the origin and is tangent to the parabola at $x = 1$. What is the equation of the line?

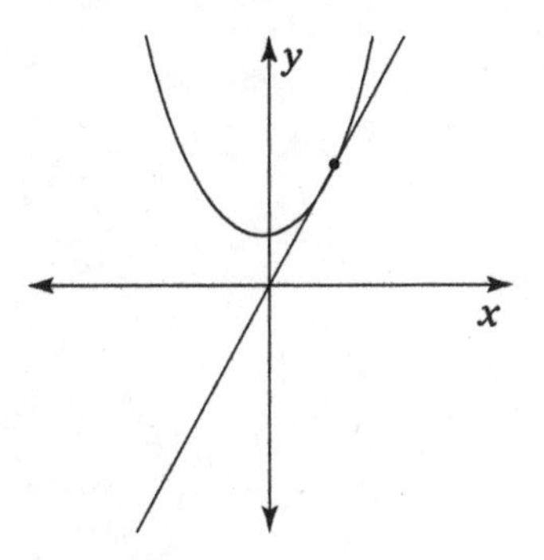

(F) $y = x$

(G) $y = 2x$

(H) $y = 2x + 1$

(J) $y = \frac{3}{2}x$

(K) $y = \frac{3}{2}x + 1$

27. In the following figure, the circle is centered at K and includes the origin. Which of the following is the formula for this circle?

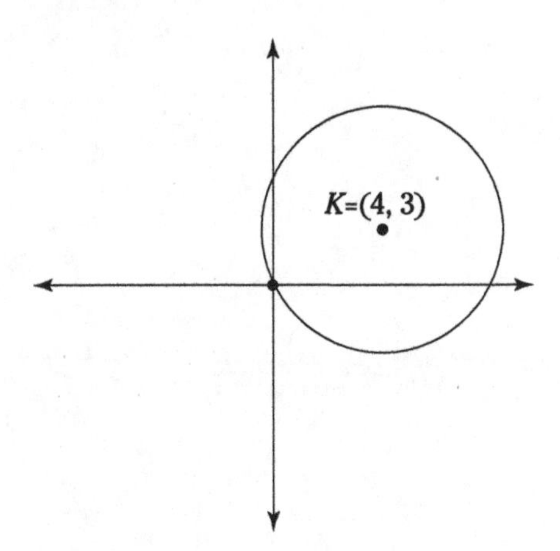

(A) $(x+4)^2 + (y+3)^2 = 5$

(B) $(x-4)^2 - (y-3)^2 = 5$

(C) $(x+4)^2 + (y+3)^2 = 25$

(D) $(x-4)^2 + (y-3)^2 = 25$

(E) $(x+4)^2 - (y+3)^2 = 25$

28. What is the range of $f(t) = \left(t^2 + 1\right)^2 - 1$?

(F) $f(t) \geq 0$

(G) $f(t) \geq 1$

(H) $f(t) \geq -1$

(J) $f(t) \geq 2$

(K) $f(t) \geq -2$

29. Which of the following graphs is a function but not a polynomial?

(A)

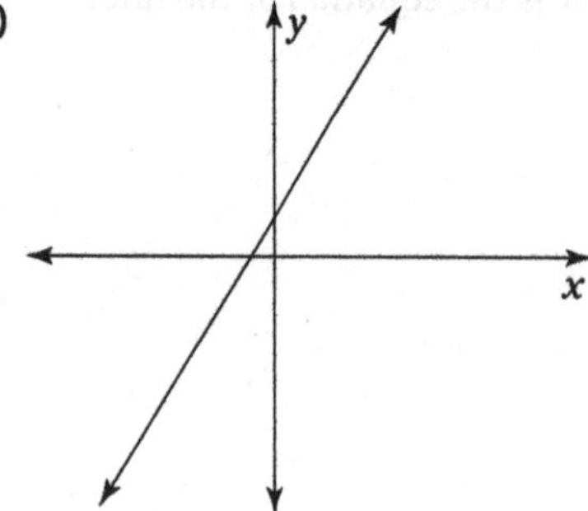

(B)

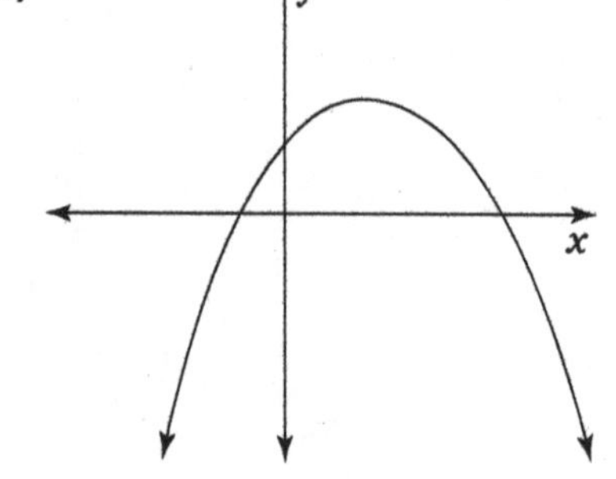

(C)

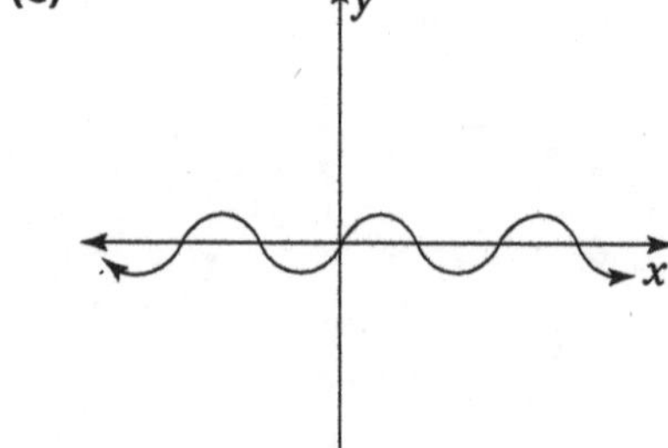

(D)

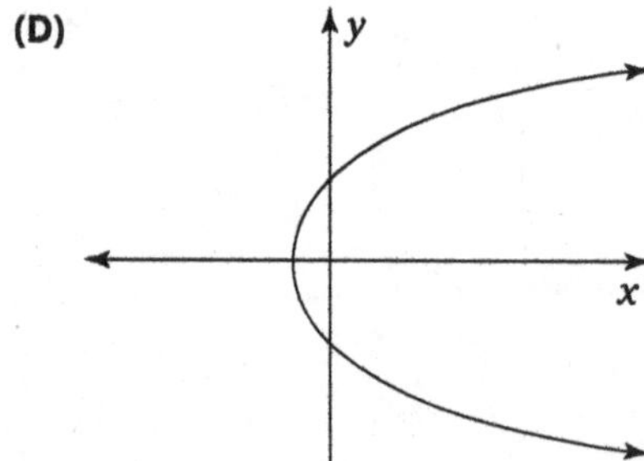

(E)

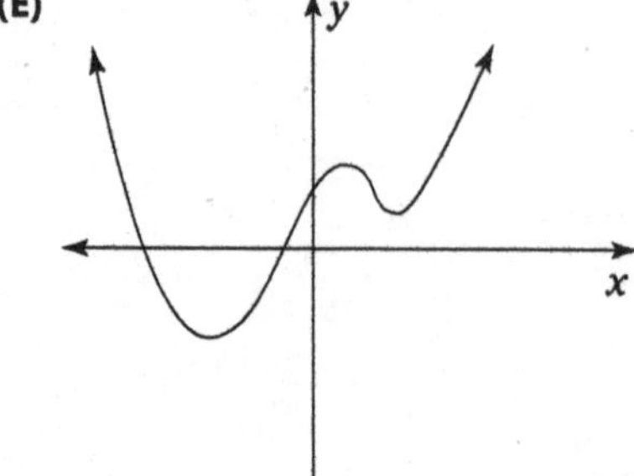

30. Patrick, Rhoda, and Tanya all work as servers at a local restaurant. This week, Patrick and Rhoda together worked 3 times as many hours as Tanya. Patrick and Tanya together worked 1.5 times as many hours as Rhoda. Rhoda and Tanya together worked 36 hours more than Patrick. What is the sum of their hours?

(F) 96

(G) 108

(H) 120

(J) 132

(K) 144

31. Given the function $f(x) = 3x^2 + 5$, what is the value of $f(a-2)$?

(A) $3a^2 + 4a - 3$

(B) $3a^2 + 4a - 7$

(C) $3a^2 + 12a - 7$

(D) $3a^2 + 12a - 12$

(E) $3a^2 - 12a + 17$

32. If q and r are the zeros of the polynomial $x^2 - 22x + 40$, then $2q + 3r =$

(F) 32 or 48

(G) 34 or 52

(H) 38 or 56

(J) 42 or 60

(K) 46 or 64

33. If $f(x) = x^2 - 4x - 10$ and $g(x) = 5x - 2$, what is the value of $f(g(3))$?

(A) 101

(B) 103

(C) 105

(D) 107

(E) 109

34. If $k > 0$ is a solution to the equation $9x^2 - 12x - 5 = 0$, then $k =$

(F) $\frac{1}{3}$

(G) $\frac{2}{3}$

(H) $\frac{4}{3}$

(J) $\frac{5}{3}$

(K) $\frac{7}{3}$

35. Which of the following is the inverse of the function $f(x)=(3x+3)^3-3$?

(A) $\frac{\sqrt[3]{x+3}+3}{3}$

(B) $\frac{\sqrt[3]{x+3}-3}{3}$

(C) $\frac{\sqrt[3]{x-3}+3}{3}$

(D) $\frac{\sqrt[3]{x-3}-3}{3}$

(E) The function has no inverse.

36. If a and b are the roots of the quadratic equation $5x^2+35x-90=0$, what is $|a-b|$?

(F) 11

(G) 15

(H) 31

(J) 45

(K) 55

Solutions to Intermediate Algebra and Coordinate Geometry Practice Problems

In this section, you find the answers to the 36 practice questions from the preceding section. I include the answers as well as the worked-out solutions.

1. **C.** First, notice that the inequality (<) is exclusive, so any equivalent inequality is also exclusive; therefore, you can rule out Choice (E). Simplify the inequality by distributing and combining like terms:

$$-5(3x-2)<25-12x$$
$$-5x+10<25-12x$$
$$-3x+10<25$$
$$-3x<15$$

Divide both sides by -3 and reverse the inequality:

$$x>-5$$

2. **K.** A negative slope slants down as the line goes from left to right, so you can rule out Choice **(F)**. Use the rise-run slope formula for each of the remaining line segments:

$$\text{Slope } \overline{AC}=\frac{\text{Rise}}{\text{Run}}=\frac{-3}{1}=-3$$
$$\text{Slope } \overline{AD}=\frac{\text{Rise}}{\text{Run}}=\frac{-5}{7}=-\frac{5}{7}$$
$$\text{Slope } \overline{BD}=\frac{\text{Rise}}{\text{Run}}=\frac{-6}{4}=-\frac{3}{2}$$
$$\text{Slope } \overline{CD}=\frac{\text{Rise}}{\text{Run}}=\frac{-2}{6}=-\frac{1}{3}$$

3. A. Let a be Anika's age and b be Bette's age. If Anika were ten years older $(a+10)$ and Bette were three years younger $(b-3)$, then Anika would be twice as old as Bette, so:

$$\begin{aligned} a+10 &= 2(b-3) \\ a+10 &= 2b-6 \\ a &= 2b-16 \end{aligned}$$

If Anika were three years older $(a+3)$ and Bette were twice her current age $(2b)$, Anika's age would be $\frac{3}{4}$ that of Bette, so:

$$\begin{aligned} a+3 &= \frac{3}{4}(2b) \\ a+3 &= \frac{3}{2}b \\ 2a+6 &= 3b \end{aligned}$$

Substitute $2b-16$ for a into the last equation:

$$\begin{aligned} 2(2b-16)+6 &= 3b \\ 4b-32+6 &= 3b \\ 4b-26 &= 3b \\ -26 &= -b \\ 26 &= b \end{aligned}$$

4. G. Begin to solve this equation by factoring:

$$\begin{aligned} x^2-9x+20 &= 0 \\ (x\quad)(x\quad) &= 0 \end{aligned}$$

List all of the combinations (including negatives) that multiply to 20:

$$\begin{array}{lll} 1\times 20 & 2\times 10 & 4\times 5 \\ -1\times -20 & -2\times -10 & -4\times -5 \end{array}$$

To factor, use the combination whose sum is –9; that is, use $-4+(-5)=-9$:

$$(x-4)(x-5)=0$$

Next, break the equation into two separate equations and solve:

$$\begin{array}{ll} x-4=0 & x-5=0 \\ x=4 & x=5 \end{array}$$

Thus, the values of a and b are 4 and 5, so

$$a^2+b^2=4^2+5^2=16+25=41$$

5. D. A line that's perpendicular to $y=5x+11$ has a slope that's the negative reciprocal of 5, so its slope is $m=-\frac{1}{5}$. The line also includes the point (0, 9), which is the y-intercept, so $b=9$. Plug these values into the slope-intercept formula to get your answer:

$$\begin{aligned} y &= mx+b \\ y &= -\frac{1}{5}x+9 \end{aligned}$$

6. H. Begin by substituting values into the functions:

$$f\left(\frac{1}{2}\right)=8\left(\frac{1}{2}\right)^2=8\left(\frac{1}{2}\right)\left(\frac{1}{2}\right)=2$$
$$g(10)=2(10)-3=20-3=17$$

Now substitute:

$$f\left(\frac{1}{2}\right)-g(10)=2-17=-15$$

7. B. $M=(-7,3)$ and $N=(5,-2)$. So, to find your answer, plug these four values into the midpoint formula:

$$\text{Midpoint}=\left(\frac{x_1+x_2}{2},\frac{y_1+y_2}{2}\right)=\left(\frac{-7+5}{2},\frac{3+(-2)}{2}\right)=\left(-1,\frac{1}{2}\right)$$

8. J. The function $g(x)$ is a reflection of $f(x)=2x^2+1$ vertically across the x-axis, so $g(x)$ equals the negation of $f(x)$:

$$g(x)=-\left(2x^2+1\right)=-2x^2-1$$

9. D. The line that includes the point (3, 2) and crosses the x-axis at $x=9$ also includes the point (9, 0). Plug these values into the two-point formula for slope:

$$\text{Slope}=\frac{y_2-y_1}{x_2-x_1}=\frac{0-2}{9-3}=\frac{-2}{6}=-\frac{1}{3}$$

10. H. The equation $3x^2+12x+4=0$ can't be solved by factoring, so use the quadratic formula:

$$x=\frac{-b\pm\sqrt{b^2-4ac}}{2a}=\frac{-12\pm\sqrt{12^2-4(3)(4)}}{2(3)}$$

Now simplify:

$$=\frac{-12\pm\sqrt{96}}{6}=\frac{-12\pm\sqrt{16}\sqrt{6}}{6}=\frac{-12\pm4\sqrt{6}}{6}=-2\pm\frac{2}{3}\sqrt{6}$$

This answer provides two possible solutions, so the correct answer is Choice (H).

11. B. The line has a slope of –2, so $m=-2$. You need to find the y-intercept, so plug in $x=7$ and $y=1$ into the slope-intercept form and solve for b:

$$y=mx+b$$
$$1=-2(7)+b$$
$$1=-14+b$$
$$15=b$$

Now plug $b=15$ and $m=-2$ back into the slope-intercept form:

$$y=-2x+15$$

12. G. When two values are inversely proportional, their product equals a constant k. If $a=3$, then $b=10$, so $ab=30$. Thus, if $a=5$, then $b=6$ (because $5\times6=30$). As a result, the correct answer is Choice (G).

13. **D.** $U=(-1,-8)$ and $V=(3,2)$. To find your answer, simply plug these values into the distance formula:

$$\text{Distance}=\sqrt{(x_2-x_1)^2+(y_2-y_1)^2}=\sqrt{(3-(-1))^2+(2-(-8))^2}$$

Now simplify:

$$=\sqrt{4^2+10^2}=\sqrt{16+100}=\sqrt{116}=\sqrt{4}\sqrt{29}=2\sqrt{29}$$

14. **F.** In the equation $y=3x^2-10x-2$, $a=3$, $b=-10$, and $c=-2$. The coefficient a is positive, so the graph is concave up, which rules out Choices (H) and (K). The coefficients a and b have different signs, which shifts the graph to the right, so Choice (J) also is ruled out. And the coefficient c is negative, which rules out Choice (G). So the only answer left is Choice (F).

15. **D.** To drop the absolute value bars, split the inequality $|5-2n|>9$ into two separate inequalities as follows:

$$5-2n>9 \qquad 5-2n<-9$$

Solve both equations for n:

$$\begin{aligned} 5-2n&>9 \qquad & 5-2n&<-9\\ -2n&>4 \qquad & -2n&<-14\\ n&<-2 \qquad & n&>7 \end{aligned}$$

16. **F.** To shift $y=x^2+2x+3$ horizontally one unit to the right, change each x to $x-1$ and simplify:

$$\begin{aligned} y&=(x-1)^2+2(x-1)+3\\ y&=(x-1)(x-1)+2(x-1)+3\\ y&=x^2-2x+1+2x-2+3\\ y&=x^2+2 \end{aligned}$$

17. **A.** The first 4 hours of service cost \$150, and each subsequent hour costs \$20. Thus, when counting hours, you need to subtract 4 hours *before* multiplying this count by 20, and then you add in \$150. Here's what the initial function will look like:

$$f(x)=20(x-4)+150$$

Then you simplify it like this:

$$\begin{aligned} f(x)&=20x-80+150\\ f(x)&=20x+70 \end{aligned}$$

18. **J.** Plug 20 for x into the function $f(x)=20x+70$, which you found in the previous problem:

$$f(20)=20(20)+70=400+70=470$$

19. **D.** To shift the function $f(x)$ units down, subtract 3 from the entire function $f(x)$, which changes it to $f(x)-3$. To shift it 6 units to the left, change x to $x+6$, which changes the function to $f(x+6)-3$. That's it. The correct answer is Choice (D).

20. **H.** In the equation $y = x^2 + 6x + 5$, $a = 1$, $b = 6$, and $c = 5$. Plug these numbers into the formula for the vertex (if you don't remember this formula, you can draw a graph or use your calculator to draw one):

$$\text{Vertex} = \left(-\frac{b}{2a}, -\frac{b^2 - 4ac}{4a}\right) = \left(-\frac{6}{2(1)}, -\frac{6^2 - 4(1)(5)}{4(1)}\right)$$

Next, simplify:

$$= \left(-\frac{6}{2}, -\frac{36 - 20}{4}\right) = (-3, -4)$$

21. **A.** Begin by treating the quadratic inequality as a quadratic equation:

$$2x^2 - x - 10 = 0$$

Next factor the equation:

$$(2x - 5)(x + 2) = 0$$

Now split it into two separate equations and solve for x:

$$2x - 5 = 0 \qquad x + 2 = 0$$
$$2x = 5 \qquad x = -2$$
$$x = \frac{5}{2}$$

The parabola for this function is concave up because the x^2 term is positive and crosses the x-axis at $\frac{5}{2}$ and –2. Because the function produces negative values between these two values of x, the inequality is satisfied between these two values. Thus, the inequality isn't satisfied when x is $-\frac{12}{5}$, making the correct answer Choice (A).

22. **H.** To answer the question, you first need to find the equation of line r. It's perpendicular to line s, whose slope is 3, so the slope of line r is $m = -\frac{1}{3}$. It intersects (0, 11), so the y-intercept of line r is $b = 11$. Plug these values into the slope-intercept form to get the equation for line r:

$$y = mx + b$$
$$y = -\frac{1}{3}x + 11$$

The equations for line r and line s form a system of equations. Set the right sides of both equations equal to each other and then solve for x:

$$-\frac{1}{3}x + 11 = 3x + 1$$
$$-x + 33 = 9x + 3$$
$$30 = 10x$$
$$3 = x$$

Now substitute 3 for x into either equation. The easier one is the equation for line s, so try that one:

$$y = 3x + 1$$
$$y = 3(3) + 1$$
$$y = 10$$

Thus, $x = 3$ and $y = 10$. So the correct answer is (3, 10).

23. A. This system of equations solves more easily by combining equations. Multiply the first equation by 5 and the second equation by 4:

$$3p + 4q = -14 \quad \text{becomes} \quad 15p + 20q = -70$$
$$7p + 5q = 15 \quad \text{becomes} \quad 28p + 20q = 60$$

Now subtract the second equation from the first:

$$\begin{aligned} 15p + 20q &= -70 \\ -28p + 20q &= 60 \\ \hline -13p &= -130 \end{aligned}$$

Solve for p:

$$p = 10$$

Finally, substitute 10 for p in one of the equations. The first equation looks easier, so go for that one:

$$\begin{aligned} 3p + 4q &= -14 \\ 3(10) + 4q &= -14 \\ 30 + 4q &= -14 \\ 4q &= -44 \\ q &= -11 \end{aligned}$$

Thus, $p + q = 10 + (-11) = -1$, so the correct answer is Choice (A).

24. K. Equation 1 is a first-degree equation, so it's linear. Equation 2 is a third-degree equation, so it's cubic. Thus, the correct answer is Choice (K).

25. B. The denominator can't equal 0. So you know that x isn't 1, because then $1 - x = 1 - 1 = 0$. Thus, Choice (A) is wrong. The value inside the square root also can't be a negative number, so:

$$\begin{aligned} 8 - x^3 &\geq 0 \\ 8 &\geq x^3 \\ 2 &\geq x \end{aligned}$$

As a result, x must be less than or equal to 2, so x can't equal 3, 4, or 5, which rules out Choices (C), (D), and (E). The correct answer has to be Choice (B).

26. G. The line is tangent to the parabola, so they share exactly one point. The x-value of this point is 1, so plug it into the equation of the parabola and solve for y:

$$y = x^2 + 1 = 1^2 + 1 = 2$$

So the coordinates of this point is (1, 2). The line includes this point and the origin (0, 0), so use the rise-run slope formula:

$$\text{Slope} = \frac{\text{Rise}}{\text{Run}} = \frac{2}{1} = 2$$

Thus, the slope of the line is $m = 2$ and its y-intercept is $b = 0$. Plug these values into the slope-intercept form to get your answer:

$$\begin{aligned} y &= mx + b \\ y &= 2x + 0 \\ y &= 2x \end{aligned}$$

27. **D.** The formula for a circle centered at (h, k) with radius r is

$$(x-h)^2+(y-k)^2=r^2$$

The circle is centered at (4, 3), so

$$(x-4)^2+(y-3)^2=r^2$$

Thus, you can rule out Choices (A), (C), and (E). The radius is the distance from (4, 3) to any point on the circle, which includes (0, 0), so use the distance formula to calculate the radius:

$$\text{Distance}=\sqrt{(x_2-x_1)^2+(y_2-y_1)^2}=\sqrt{(4-0)^2+(3-0)^2}$$

Now simplify:

$$=\sqrt{4^2+3^2}=\sqrt{16+9}=\sqrt{25}=5$$

The radius is $r=5$, so the formula is as follows:

$$(x-4)^2+(y-3)^2=5^2$$
$$(x-4)^2+(y-3)^2=25$$

28. **F.** Start from the inside of the function $f(t)=(t^2+1)^2-1$ and work your way out: The value of t^2 can't be negative, so its lowest value is 0; therefore, the range of $f(t)=t^2$ is $f(t)\geq 0$. So when you add 1 to this function, its lowest value becomes 1; therefore, the range of $f(t)=t^2+1$ is $f(t)\geq 1$. When you square this function, its lowest value remains 1; therefore, the range of $f(t)=t^2+1$ is $f(t)\geq 1$. Finally, when you subtract 1 from this function, its lowest value becomes 0; therefore, the range of $f(t)=(t^2+1)^2-1$ is $f(t)\geq 1$. To test this, notice that when $t=0$, $f(t)=0$, but at all other values of t, $f(t)>0$.

29. **C.** Choice (D) isn't a function, because at least one x-value has more than one y-value (that is, a vertical line can pass through it twice). So you can rule that choice out. Choices (A), (B), and (E) are all polynomials — linear, quadratic, and fourth-degree, respectively — so you can rule them out as well. The correct answer is Choice (C), which is a function but not a polynomial.

30. **H.** Let p, r, and t, respectively, be the number of hours that Patrick, Rhoda, and Tanya worked. With those variables you can make these three equations:

$$p+r=3t$$
$$p+t=1.5r$$
$$r+t=p+36$$

Subtract r from both sides of the first equation:

$$p=3t-r$$

Now substitute $3t-r$ for p into the second and third equations:

$$3t-r+t=1.5r$$
$$r+t=3t-r+36$$

Combine like terms in these two equations:

$$4t=2.5r$$
$$2r=2t+36$$

You can further simplify these by multiplying the first equation by 2 and dividing the second equation by 2:

$$8t = 5r$$
$$r = t + 18$$

Substitute $t + 18$ for r in the first equation and solve for t:

$$8t = 5(t + 18)$$
$$8t = 5t + 90$$
$$3t = 90$$
$$t = 30$$

Substitute 30 for t back into one of the two previous equations ($r = t + 18$ looks easiest) and solve for r:

$$r = 30 + 18$$
$$r = 48$$

Finally, substitute 30 for t and 48 for r into any of the original equations ($p + r = 3t$ looks easiest) and solve for p:

$$p + 48 = 3(30)$$
$$p + 48 = 90$$
$$p = 42$$

Thus, the sum of their hours is $p + r + t = 42 + 48 + 30 = 120$, making Choice (H) the correct answer.

31. **E.** Begin by making the following substitution:

$$f(a-2) = 3(a-2)^2 + 5 = 3(a-2)(a-2) + 5$$

Complete the problem by FOILing and then combining like terms:

$$= 3a^2 - 12a + 12 + 5 = 3a^2 - 12a + 17$$

32. **K.** To begin, set the expression $x^2 - 22x + 40$ to 0 and solve for x:

$$x^2 - 22x + 40 = 0$$
$$(x-2)(x-20) = 0$$
$$x = 2,\ 20$$

Thus, q and r equal, in some order, 2 and 20, so:

$2q + 3r = 2(20) + 3(2) = 46$ or $2q + 3r = 2(2) + 3(20) = 64$

33. **D.** First, use the function $g(x) = 5x - 2$ to find $g(3)$:

$$g(3) = 5(3) - 2 = 13$$

Thus, $f(g(3)) = f(13)$, so use the function $f(x) = x^2 - 4x - 10$ to find $f(13)$:

$$f(13) = 13^2 - 4(13) - 10 = 169 - 52 - 10 = 107$$

34. **J.** To solve the equation, you need to split $12x$ into two equivalent terms whose coefficients multiply to $ac = -45$ and add up to $b = -12$. These values are 3 and -15:

$$9x^2 - 12x - 5 = 0$$
$$9x^2 + 3x - 15x - 5 = 0$$

Now, factor the left side of this equation by grouping:

$$3x(3x+1) - 5(3x+1) = 0$$
$$(3x-5)(3x+1) = 0$$

Solve for both values of x:

$$3x - 5 = 0 \qquad 3x + 1 = 0$$
$$3x = 5 \qquad 3x = -1$$
$$x = \frac{5}{3} \qquad x = -\frac{1}{3}$$

Thus, the positive solution is $\frac{5}{3}$.

35. **B.** Begin by setting the function to y:

$$y = (3x+3)^3 - 3$$

Next, then exchange the variables x and y:

$$x = (3y+3)^3 - 3$$

Solve for y:

$$x + 3 = (3y+3)^3$$
$$\sqrt[3]{x+3} = 3y + 3$$
$$\sqrt[3]{x+3} - 3 = 3y$$
$$\frac{\sqrt[3]{x+3} - 3}{3} = y$$

36. **F.** Begin solving the equation by dividing out the GCF:

$$5x^2 + 35x - 90 = 0$$
$$5\left(x^2 + 7x - 18\right) = 0$$
$$x^2 + 7x - 18 = 0$$

Factor and solve:

$$x^2 + 7x - 18 = 0$$
$$(x+9)(x-2) = 0$$
$$x = -9,\ 2$$

Set these values to a and b in any order:

$$|a - b| = |-9 - 2| = |-11| = 11$$

4 Visualizing Plane Geometry and Trigonometry

IN THIS PART . . .

Understanding lines and angles, triangles, quadrilaterals, circles, and solids.

Working with trig ratios, radian measure, matrices, logarithms, and imaginary numbers.

Solving practice problems for all of these topics.

IN THIS CHAPTER

» Answering questions involving angles

» Familiarizing yourself with triangles

» Making peace with quadrilaterals

» Discovering the ins and outs of circles

» Looking into solid geometry

Chapter 11 Plain Talk about Plane Geometry

In this chapter, the topic is geometry. Although a typical geometry course spends a lot of time focusing on geometric proofs, what you need to succeed on the ACT is a review of the basics.

The chapter begins with a refresher on angles, including vertical angles, supplementary angles, angles that involve parallel lines, and angles in a polygon. Then I move on to triangles. I discuss the area formula for a triangle, the Pythagorean theorem, and a variety of common right triangles that you're sure to see on the ACT.

Next, I move on to quadrilaterals, showing you the formulas for the area and perimeter of squares and rectangles as well as the area of a parallelogram and a trapezoid. I also show you how to find the area and perimeter of a circle and how to answer ACT questions that involve tangent lines, arc length, and chords.

I round out the chapter with a discussion of *solid geometry*, which is the extension of plane geometry to three-dimensional solids. I introduce you to the formulas for the volume and surface area of a cube, a box (rectangular solid), and a sphere. I also give you a few formulas for a variety of other common solids.

Knowing Your Angles

In your geometry class, you probably spent a lot of time proving theorems about angles. This is good training for clear thinking and, of course, for passing your geometry tests. But, luckily, on the ACT, you won't have to write any geometric proofs. However, you do need to know how to work angles. So, in this section, I give you the most important rules about angles and show you how to work with them to answer a variety of typical ACT questions.

Angles around one point

Here are three important rules for answering ACT questions about angles:

- All the angles around a single point add up to 360°.
- *Supplementary angles* — any pair of angles that combine to produce a straight line — add up to 180° degrees. (The angles on both sides of a line add up to the total angles around a single point, because $180° + 180° = 360°$.)
- *Vertical angles* — any pair of opposite angles formed when two lines intersect — are equal to each other. (Vertical angles are symmetrical to each other, so they're equal.)

I show these relationships among angles in Figure 11-1.

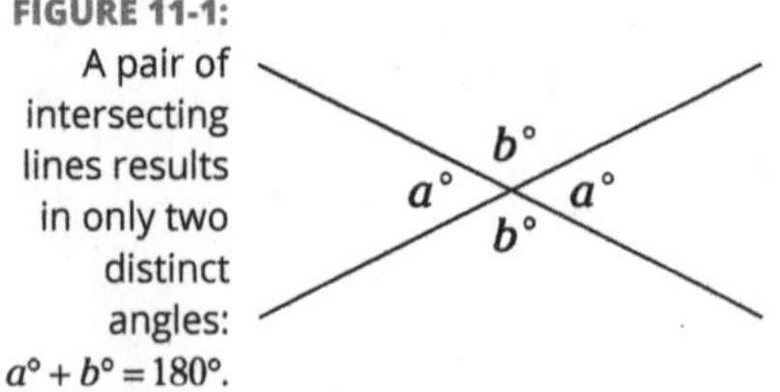

FIGURE 11-1: A pair of intersecting lines results in only two distinct angles: $a° + b° = 180°$.

EXAMPLE

In the following figure, what is the value of m?

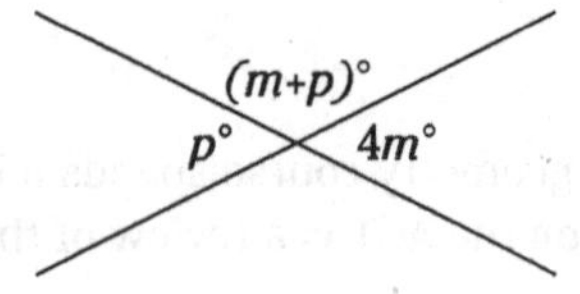

(A) 10

(B) 12

(C) 15

(D) 18

(E) 20

The figure shows that $p°$ and $4m°$ are a pair of vertical angles, so

$$p = 4m$$

Additionally, $p°$ and $(m+p)°$ are supplementary angles, so

$$p + (m+p) = 180$$

Simplify the second equation:

$$2p + m = 180$$

Substitute $4m$ for p and solve:

$$\begin{aligned} 2(4m) + m &= 180 \\ 8m + m &= 180 \\ 9m &= 180 \\ m &= 20 \end{aligned}$$

Therefore, the correct answer is Choice (E).

Angles and parallel lines

A pair of *parallel lines* extends infinitely in both directions at a constant distance from each other. Geometry includes a bunch of theorems about parallel lines. But don't fret. For the ACT, you only need to be able to work with the angles that result when a line crosses a pair of parallel lines, as shown in Figure 11-2.

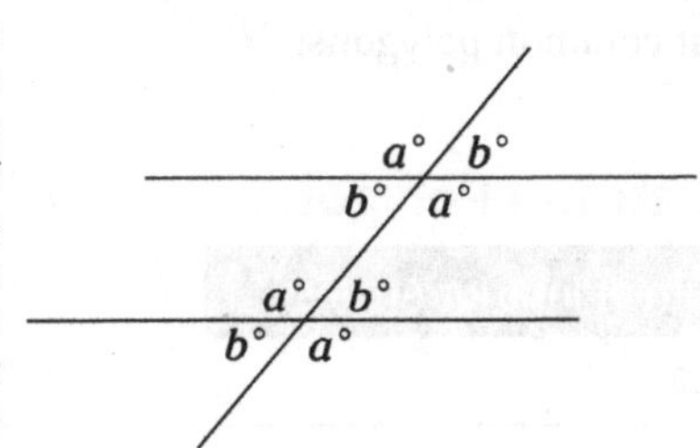

FIGURE 11-2: A line intersecting a pair of parallel lines results in only two distinct angles: $a° + b° = 180°$.

EXAMPLE

In the following figure, the two horizontal lines are parallel. What is the value of $h + j$?

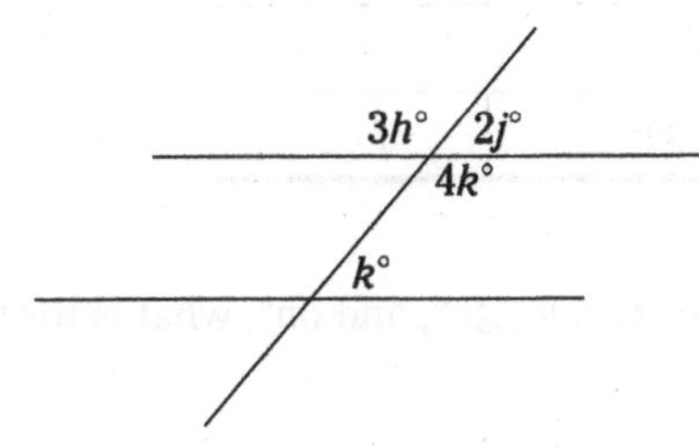

(A) 56

(B) 66

(C) 76

(D) 78

(E) 80

Because the two lines are parallel, the preceding figure has only two distinct angles. The smaller angle measures $k°$ and the larger angle measures $4k°$. These two angles add up to 180°, so you can create and solve the following equation:

$$k + 4k = 180$$
$$5k = 180$$
$$k = 36$$

The smaller angle measures 36°, so

$$2j = 36$$
$$j = 18$$

Additionally, the larger angle measures 144° (because $36 + 144 = 180$), so

$$3h = 144$$
$$h = 48$$

Therefore,

$$h + j = 48 + 18 = 66$$

So the correct answer is Choice (B).

Interior angles in a polygon

A *polygon* is a shape made entirely of straight line segments. When you know how many sides a polygon has, you can find the sum of the *interior angles* for that polygon — that is, the total of all angles inside the polygon — using the following formula:

$$\text{Total interior angles} = 180(\text{Number of angles} - 2)$$

Table 11-1 gives you the sum of angles for the most common polygons.

TABLE 11-1: Sum of Interior Angles of Common Polygons

Polygon	Number of Sides	Sum of Interior Angles
Triangle	3	180°
Quadrilateral	4	360°
Pentagon	5	540°
Hexagon	6	720°
Heptagon	7	900°
Octagon	8	1,080°

EXAMPLE

If the measures of four angles in a quadrilateral are 140°, n°, $3n$°, and $6n$°, what is the value of n?

(A) 10

(B) 20

(C) 22

(D) 24

(E) 36

The total measure of the four angles is 360°, so set up the following equation:

$$140 + n + 3n + 6n = 360$$

Now solve for n:

$$\begin{aligned} 140 + 10n &= 360 \\ 10n &= 220 \\ n &= 22 \end{aligned}$$

Thus, the correct answer is Choice (C).

Working with Triangles

A *triangle* is a shape with three sides that are line segments. These three-sided shapes are tremendously important in geometry and are the basis of *trigonometry* (the study of triangles), which I discuss in Chapter 12. Here are three important types of triangles:

- An *equilateral* triangle has three equal sides and three equal angles all measuring 60°.

- An *isosceles* triangle has two equal sides and two corresponding angles that are equal to each other.
- A *right* triangle has one right angle — that is, one 90° angle.

In this section, you get a solid basis for answering ACT questions that include triangles. First, I introduce the formula for the area of a triangle. Then, I discuss a set of special triangles with distinct properties. Next, I move on to the Pythagorean theorem, which allows you to calculate the length of one side of a right triangle when you know the lengths of the other two sides.

Finding the area of a triangle

To find the area of any triangle, you need to know the *base* (b), which can be any side, and its *height* (h), which is the distance from the base to the opposite corner. I illustrate this concept in Figure 11-3. Here's the formula for the area of a triangle:

$$A = \frac{1}{2}bh$$

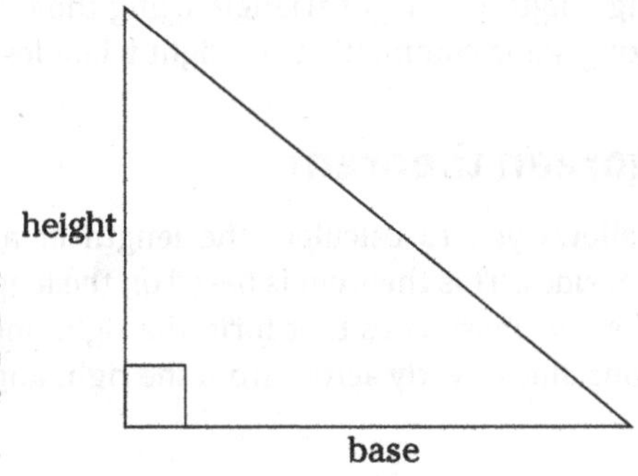

FIGURE 11-3: The formula for the area of a triangle is $A = \frac{1}{2}bh$.

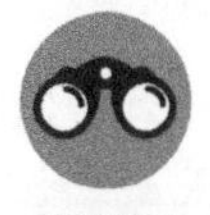

EXAMPLE

In the following figure, if the area of ΔPQT is 14, what is the area of ΔPQR?

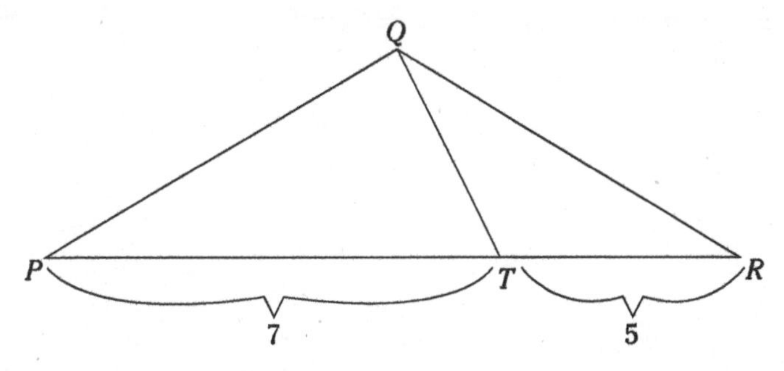

(A) 10

(B) 20

(C) 24

(D) 28

(E) 48

You know the area and base of ΔPQT, so use the formula for the area of a triangle to find the height:

$$A = \frac{1}{2}bh$$

$$14 = \frac{1}{2}(7)h$$

$$28 = 7h$$

$$4 = h$$

The height of ΔPQT is 4, so the height of ΔPQR is also 4. The base of ΔPQR is 12, so

$$A = \frac{1}{2}(12)(4) = 24$$

So the correct answer is Choice (C).

Answering questions containing right triangles

A *right triangle* is a triangle that has one right angle — an angle that measures 90°. Right triangles appear often on the ACT, so you need to know how to work with them and recognize them. In this section, I show you how to calculate a missing length of a right triangle using the Pythagorean theorem. I also provide some info on recognizing some commonly used right triangles.

Taking advantage of the Pythagorean theorem

The *Pythagorean theorem* is a formula that allows you to calculate the length of a side of a right triangle using the lengths of the other two sides. This theorem is based on the lengths of the two *legs* (a and b) of the triangle — that is, the two short sides that form the right angle — and the length of the *hypotenuse* (c), which is the long side directly across from the right angle. Here's what the theorem looks like:

$$a^2 + b^2 = c^2$$

EXAMPLE

What is the area of the following right triangle?

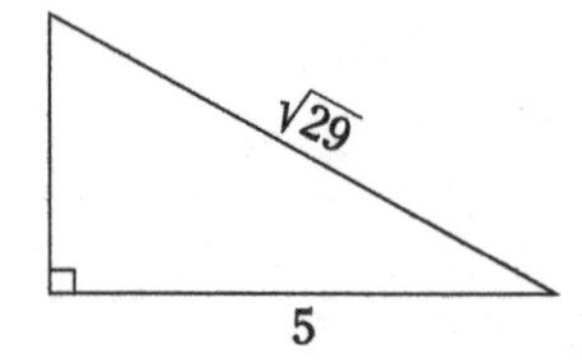

(A) 2

(B) 4

(C) 5

(D) 8

(E) 10

You know the length of one leg (5) and the hypotenuse ($\sqrt{29}$), so plug these values into the formula for the Pythagorean theorem:

$$\begin{aligned} a^2 + b^2 &= c^2 \\ 5^2 + b^2 &= \sqrt{29}^2 \\ 25 + b^2 &= 29 \\ b^2 &= 4 \\ b &= 2 \end{aligned}$$

As you can see, the remaining leg of the triangle has a length of 2. In a right triangle, the two legs are also the base and height, so plug these two values into the formula for the area of a triangle:

$$A = \frac{1}{2}bh = \frac{1}{2}(2)(5) = 5$$

Therefore, the correct answer is Choice (C).

Recognizing some common right triangles

Two of the most common right triangles on the ACT are identified by the measurements of their angles: the 45-45-90 triangle and the 30-60-90 triangle. You can see examples of these two triangles in Figure 11-4.

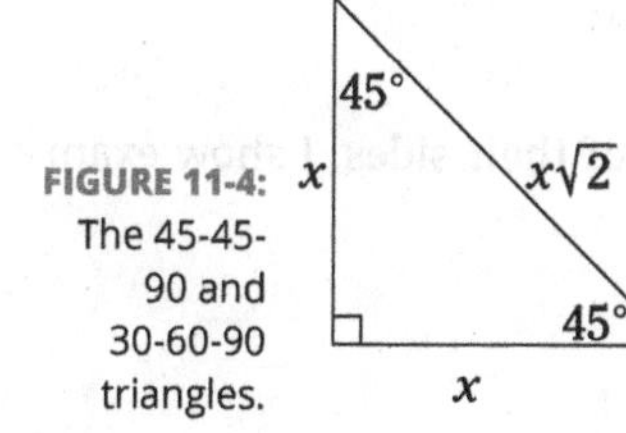

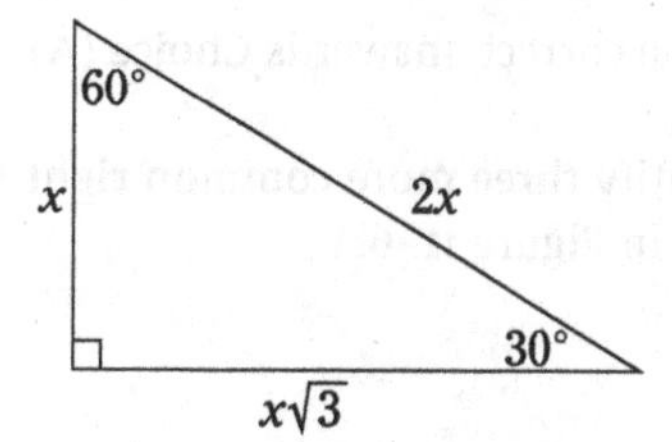

FIGURE 11-4: The 45-45-90 and 30-60-90 triangles.

The 45-45-90 triangle has three sides whose ratios are of these lengths: $x : x : x\sqrt{2}$. The 30-60-90 triangle has three sides whose ratios are of these lengths: $x : x\sqrt{3} : 2x$.

TIP

You can put these two triangles to use when trying to answer a wide variety of ACT questions that seem unrelated to right triangles. For example, notice that when you bisect a square diagonally, the result is two 45-45-90 triangles. Similarly, when you bisect an equilateral triangle vertically, the result is two 30-60-90 triangles. Check out Figure 11-5 to see these bisections and resulting triangles.

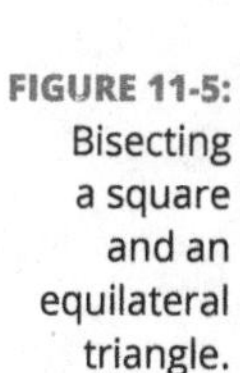

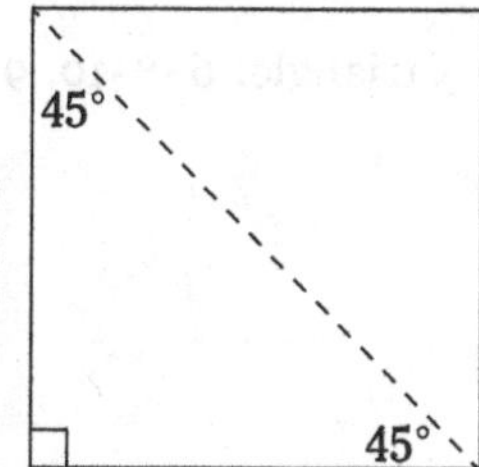

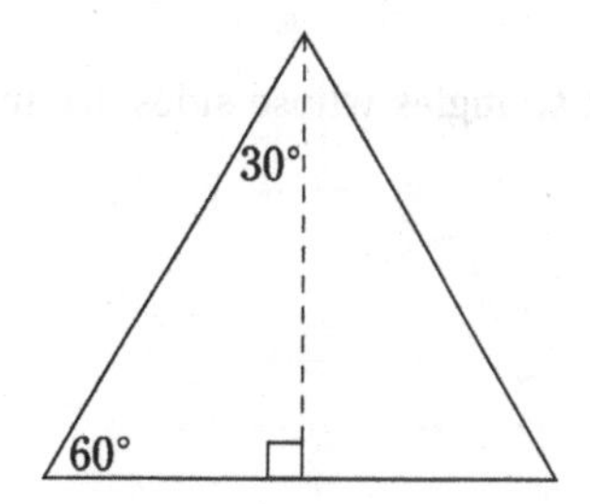

FIGURE 11-5: Bisecting a square and an equilateral triangle.

EXAMPLE

In the following figure, what is the ratio of FN to GH?

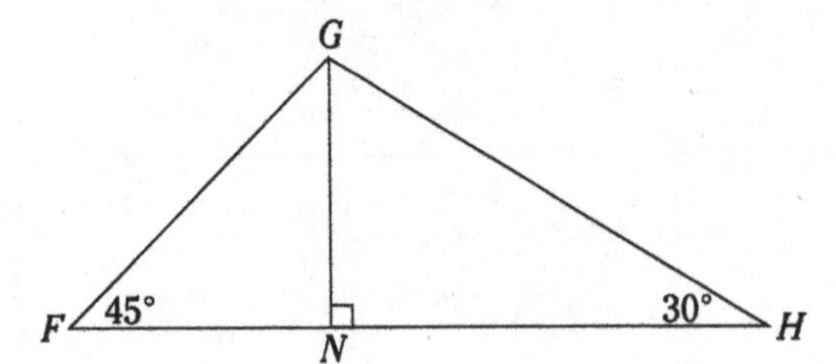

(A) 1 to 2

(B) 1 to $\sqrt{2}$

(C) 1 to $\sqrt{3}$

(D) 2 to $\sqrt{3}$

(E) $\sqrt{2}$ to $\sqrt{3}$

Angle F measures 45°, so ΔGNF is a 45-45-90 triangle. Similarly, $\angle H$ measures 30°, so ΔHNG is a 30-60-90 triangle. Let $FN = x$. Then, $GN = x$ and $GH = 2x$, and you can calculate the ratio of FN to GH like this:

$$\frac{FN}{GH} = \frac{x}{2x} = \frac{1}{2}$$

Therefore, the correct answer is Choice (A).

You can identify three more common right triangles by the lengths of their sides. I show examples of them in Figure 11-6.

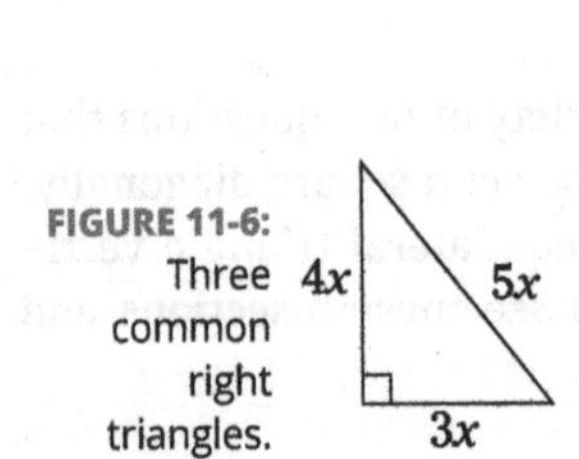

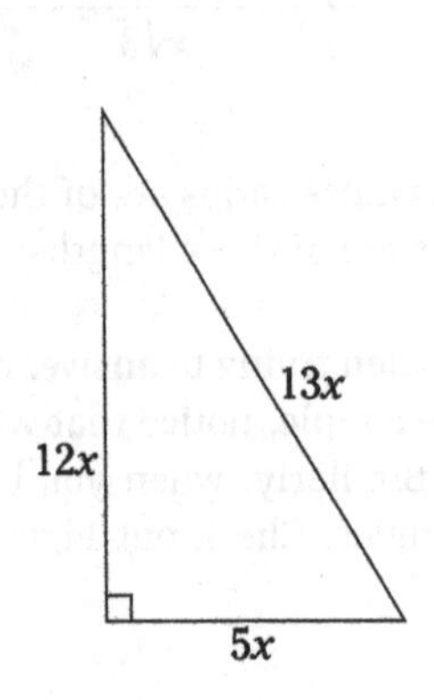

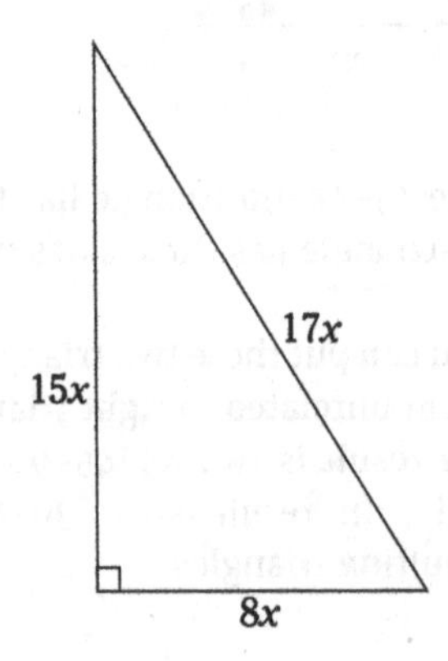

FIGURE 11-6: Three common right triangles.

TIP

Look for right triangles whose sides are multiples of the 3-4-5 triangle: 6-8-10, 9-12-15, and so on.

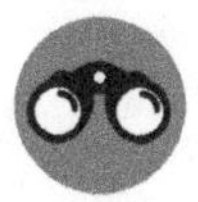

In the following figure, what is the value of x?

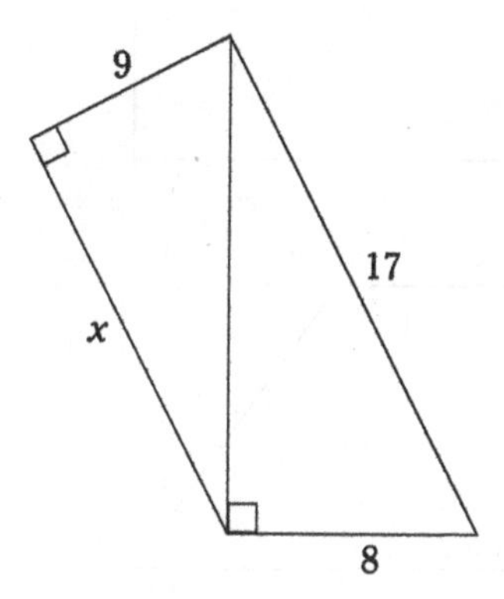

(A) 9

(B) 10

(C) 12

(D) 13

(E) 15

The figure shows two right triangles. The triangle on the right side of the figure has a leg that is the length of 8 and a hypotenuse that is the length of 17, so it's an 8-15-17 triangle; therefore, the length of its remaining leg is 15. The triangle on the left has a leg with the length of 9 and a hypotenuse the length of 15, so it's the 9-12-15 variation of the 3-4-5 triangle. Therefore, $x = 12$, making the correct answer Choice (C).

Breezing through Your Work with Quadrilaterals

A *quadrilateral* is a shape with four sides that are line segments. For the ACT, you need to be able to work with the following four basic quadrilaterals:

- **Squares:** A *square* has four right angles and four sides that are equal in length.
- **Rectangles:** A *rectangle* has four right angles and two pairs of opposite sides that are equal in length.
- **Parallelograms:** A *parallelogram* has two pairs of opposite sides that are parallel to each other and equal in length.
- **Trapezoids:** A *trapezoid* has one pair of opposite sides that are parallel to each other.

Figure 11-7 shows you these four basic quadrilaterals.

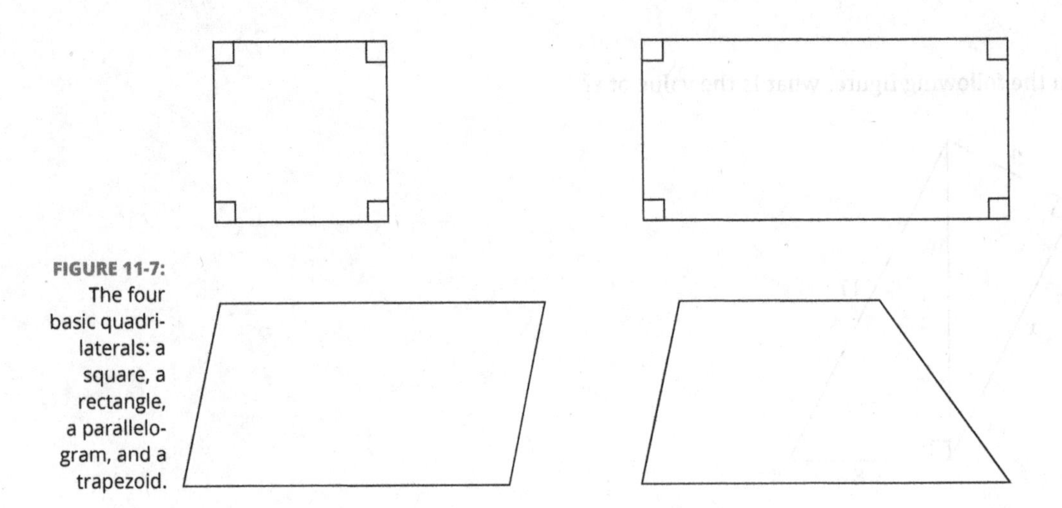

FIGURE 11-7: The four basic quadrilaterals: a square, a rectangle, a parallelogram, and a trapezoid.

In this section, I get you up to speed on a variety of questions that cover these four types of quadrilaterals.

Squares

A *square* is a quadrilateral with four equal sides and four right angles. The area and perimeter of a square are both based on the length of one side (s):

- **Area:** $A = s^2$
- **Perimeter:** $P = 4s$

EXAMPLE

If the perimeter of a square is $8n$ and its area is $20n$, what is the value of n?

(A) 2

(B) 4

(C) 5

(D) 8

(E) 10

The perimeter of the square is $8n$, so you can use the perimeter formula to express the length of a side (s) in terms of n:

$$P = 4s$$
$$8n = 4s$$
$$2n = s$$

Now plug this value of s into the area formula:

$$A = s^2 = (2n)^2 = 4n^2$$

The question tells you that the area is also $20n$, so:

$$4n^2 = 20n$$
$$n^2 = 5n$$
$$n = 5$$

Therefore, the correct answer is Choice (C).

Rectangles

A *rectangle* is a quadrilateral with four right angles and two pairs of equal sides. The area and perimeter of a rectangle are both based on its *length* (l), which is its longer side, and its *width* (w), which is its shorter side:

- **Area:** $A = lw$
- **Perimeter:** $P = 2l + 2w$

EXAMPLE

A rectangular field has a length and width that are in a ratio of 4:3. If the diagonal distance from one corner of the field to the opposite corner is 200 feet, what is the perimeter of the field?

(A) 360 feet

(B) 480 feet

(C) 540 feet

(D) 560 feet

(E) 640 feet

Begin by drawing a picture like the following to get a sense of what the question asks you for:

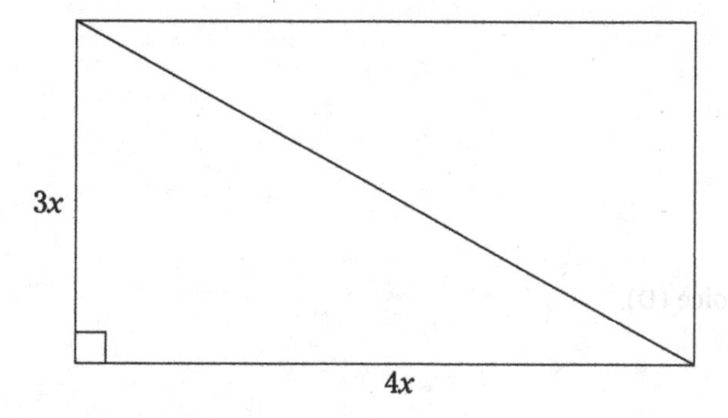

TIP

This right triangle has two legs that are $3x$ and $4x$, so it's a 3-4-5 triangle. If you notice this fact, you can see that the hypotenuse is $5x$. As a result, you can set up the equation $5x = 200$, which solves as $x = 40$.

But even if you don't notice this fact, you can still see that the right triangle created by the diagonal has two legs that are labeled $3x$ and $4x$ (using the ratio) with the diagonal being 200. So it's easy to set up for the Pythagorean theorem:

$$\begin{aligned}(3x)^2 + (4x)^2 &= 200^2 \\ 9x^2 + 16x^2 &= 40{,}000 \\ 25x^2 &= 40{,}000 \\ x^2 &= 1{,}600 \\ x &= 40\end{aligned}$$

Thus, the length of the field is $4 \times 40 = 160$ and the width is $3 \times 40 = 120$. Plug these values into the formula for the perimeter of a rectangle:

$$A = 2l + 2w = 2(160) + 2(120) = 320 + 240 = 560$$

So the correct answer is Choice (D).

Parallelograms

A *parallelogram* is a quadrilateral with two pairs of parallel sides. No special formula exists for finding the perimeter of a parallelogram. But, to find the area of one, you need to know the *base* (*b*), which can be any side, and its *height* (*h*), which is the shortest distance from the base to the opposite side:

$$A = bh$$

EXAMPLE

The height of a parallelogram is n and the length of its base is $2.5n$. What is the value of n if the area of the parallelogram is 20?

(A) 1

(B) 2

(C) $\sqrt{2}$

(D) $2\sqrt{2}$

(E) $\frac{\sqrt{2}}{2}$

Plug the height, base, and area into the formula for the area of a parallelogram:

$$A = bh$$
$$20 = 2.5n(n)$$
$$20 = 2.5n^2$$
$$8 = n^2$$
$$\sqrt{8} = n$$
$$2\sqrt{2} = n$$

Thus, the correct answer is Choice (D).

Trapezoids

A *trapezoid* is a quadrilateral with one pair of parallel sides. No special formula exists for finding the perimeter of a trapezoid. You simply add up the side lengths. The area of this quadrilateral depends upon the *bases* (b_1 *and* b_2), which are the two parallel sides, and its *height* (*h*), which is the distance between the two bases. Here's what the formula for the area looks like:

$$A = \frac{b_1 + b_2}{2}h$$

TIP

If you need help remembering this formula, notice that the fractional part is the average of the two bases. This average multiplied by the height of the trapezoid gives you the area.

EXAMPLE

In the following figure, the trapezoid *GHJK* has an area of 25. Which of the following is the approximate whole-number distance from *H* to *K*?

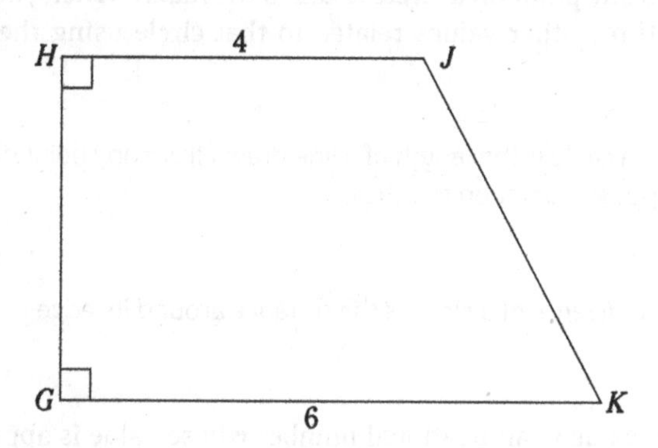

(A) 7

(B) 8

(C) 9

(D) 10

(E) 11

The figure shows that the lengths of the two bases are 4 and 6, and the area of the trapezoid is 25, so use the area formula for a trapezoid to find the height:

$$A = \frac{b_1 + b_2}{2}h$$

$$25 = \frac{4+6}{2}h$$

$$25 = 5h$$

$$5 = h$$

Thus, HG = 5. The distance from *H* to *K* is the hypotenuse of ΔGHK. The lengths of the two sides of this triangle are 5 and 6, so use the Pythagorean theorem to find the hypotenuse:

$$c^2 = a^2 + b^2 = 5^2 + 6^2 = 25 + 36 = 61$$

Because $c = \sqrt{61} \approx 7.8$, the correct answer is Choice (B).

Wheeling and Dealing with Circles

A *circle* is the set of points that are all an equal distance from a single center point. ACT questions often require you to find values related to a circle, such as its diameter, area, or circumference. Knowing how to work with a line tangent to a circle is also helpful. Finally, you need to know a bit about arc length. In this section, I help you roll through all these concepts.

Rounding up the basic circle formulas

The distance from the center point to any point on a circle is called the *radius*. When you know the radius (r) of a circle, you can find three other values related to that circle using the following formulas:

- **Diameter:** $D = 2r$. The *diameter* of a circle is the length of a line drawn from any point on the circle through the center to the opposite point on the circle.
- **Area:** $A = \pi r^2$.
- **Circumference:** $C = 2\pi r$. The *circumference* of a circle is the distance around its edge — essentially, its perimeter.

REMEMBER

Two of these three formulas use the value π, an irrational number whose value is approximately 3.14. However, an ACT question won't ask you to substitute a numerical value for π without specifying what value to use.

EXAMPLE

The area of a circle is n and its circumference is also n. What is its diameter?

(A) 2

(B) $2n$

(C) $2\pi n$

(D) 4

(E) $4\pi n$

The area of the circle is n, so

$$n = \pi r^2$$

The circumference is also n, so

$$n = 2\pi r$$

Because both values equal n, you can set them equal to each other and solve for r:

$$2\pi r = \pi r^2$$
$$2r = r^2$$
$$2 = r$$

Now use the radius to find the diameter:

$$D = 2r = 2 \times 2 = 4$$

So the correct answer is Choice (D).

Understanding tangent lines

A *tangent line* is a line that touches a circle at a single point, as shown in Figure 11-8. The most important feature of a tangent line is that it forms a right angle with the radius (or diameter) that also crosses that point. Knowing this fact is essential to answering questions on the ACT.

WARNING

Don't confuse a tangent line to a circle with the concept of a tangent in trigonometry (which I discuss in Chapter 12). You don't need trig to answer an ACT question about tangent lines.

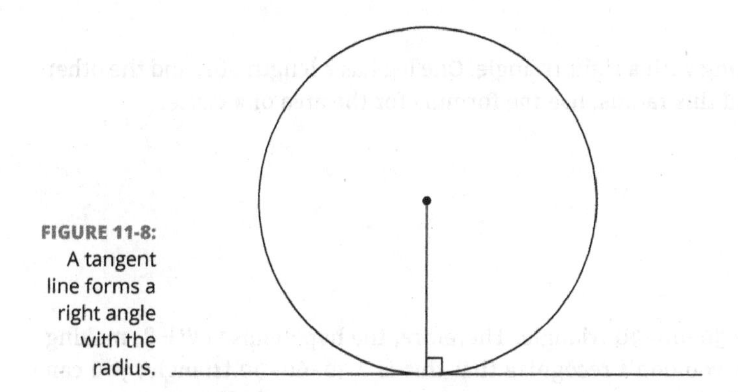

FIGURE 11-8: A tangent line forms a right angle with the radius.

When an ACT question includes a circle with a tangent line, look for a right triangle, and then see whether it's one of the three common right triangles I discuss in the earlier section "Recognizing some common right triangles." Even if it isn't, you still may be able to use the Pythagorean theorem to answer the question.

EXAMPLE

In the following figure, the circle has an area of 3π, and $\overline{AB}$ is a segment of a tangent line that has a length of 1. What is the length of $\overline{OB}$?

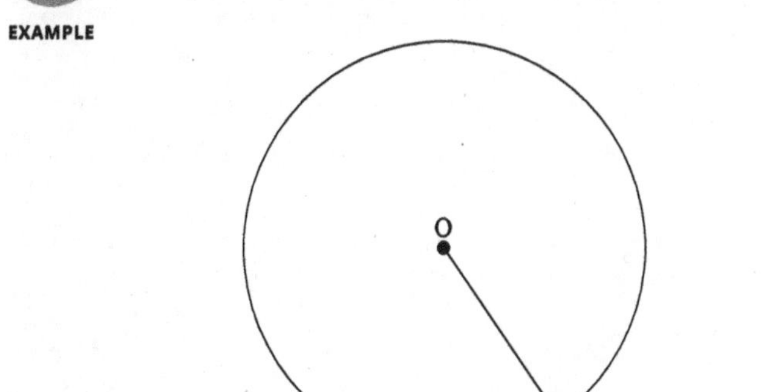

(A) 2

(B) 2π

(C) $\sqrt{2}$

(D) 3

(E) $\sqrt{3}$

The key to this question is to add $\overline{OA}$ to the figure like this:

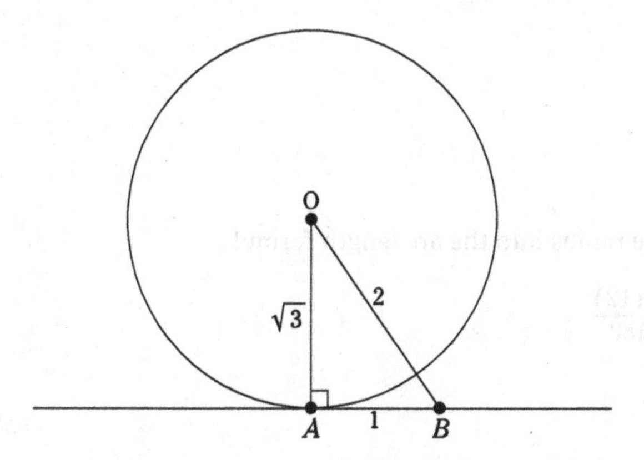

Now you can see that you're working with a right triangle. One leg has a length of 1, and the other leg is a radius of the circle. To find this radius, use the formula for the area of a circle:

$$A = \pi r^2$$
$$3\pi = \pi r^2$$
$$3 = r^2$$
$$\sqrt{3} = r$$

Because $OA = \sqrt{3}$, the triangle is a 30-60-90 triangle. Therefore, the hypotenuse $OB = 2$, making the correct answer Choice (A). (If you don't recognize that this is a 30-60-90 triangle, you can still find the solution using the Pythagorean theorem. However, keep in mind that it's a more complex calculation.)

Making sense of arc length

Arc length is a partial distance around the edge of a circle. For example, Figure 11-9 shows a circle with a 45° arc highlighted.

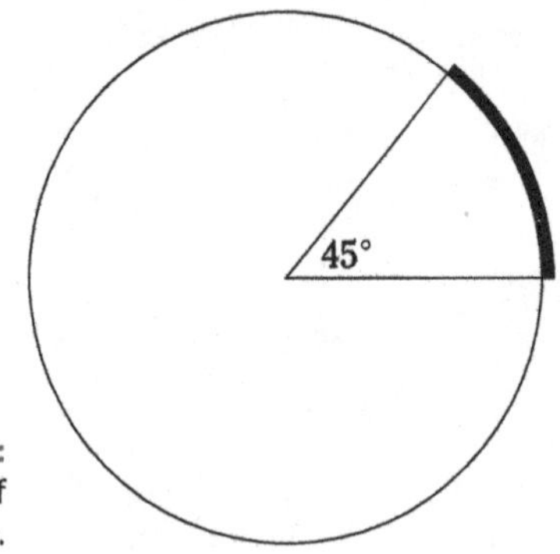

FIGURE 11-9: A 45° arc of a circle.

Arc length makes sense in relation to circumference of a circle, $C = 2\pi r$. In fact, the circumference of a circle is simply a 360° arc length. With that understanding, the arc length formula becomes easier to handle:

$$\text{Arc length} = \text{degrees}\,\frac{\pi r}{180}$$

EXAMPLE

What is the length of a 60° arc of a circle with a radius of 12?

(A) π

(B) 2π

(C) 3π

(D) 4π

(E) 6π

Plug the degrees of the arc and the radius into the arc length formula:

$$\text{Arc length} = \text{degrees}\,\frac{\pi r}{180} = 60\frac{\pi(12)}{180}$$

Now simplify:

$$= \frac{\pi(12)}{3} = 4\pi$$

Thus, the correct answer is Choice (D).

EXAMPLE

If a 40° arc of a circle has a length of $\frac{10\pi}{9}$, what is the area of the circle?

(F) π

(G) 9π

(H) 25π

(J) 36π

(K) 81π

Begin by plugging the degrees and arc length into the formula:

$$\text{Arc length} = \text{degrees}\,\frac{\pi r}{180}$$

$$\frac{10\pi}{9} = 40\frac{\pi r}{180}$$

Next, simplify and solve for r:

$$\frac{10\pi}{9} = \frac{2\pi r}{9}$$

$$10\pi = 2\pi r$$

$$5 = r$$

Now plug r into the area formula for a circle:

$$A = \pi r^2 = \pi(5)^2 = 25\pi$$

So the correct answer is Choice (H).

Striking a few chords

A *chord* of a circle is a line segment drawn between two points on that circle. See Figure 11-10 for an example of a chord.

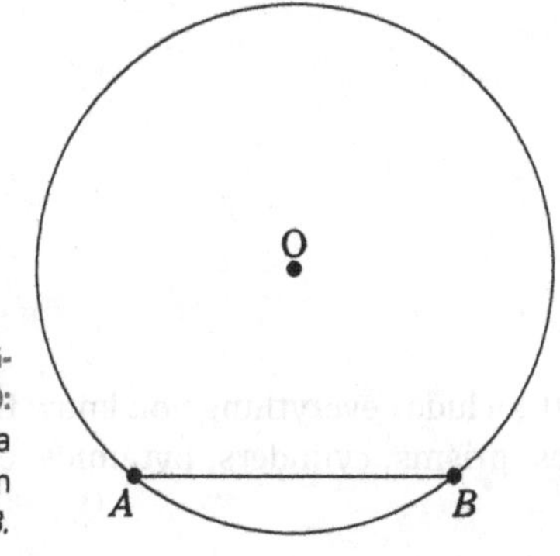

FIGURE 11-10: A chord of a circle from *A* to *B*.

TIP

A line drawn from the center of a circle to the nearest point on the chord bisects the chord at a right angle. This fact can be useful for answering ACT questions, as the following example illustrates.

EXAMPLE

P and Q are two points on a circle with a center named O, such that $\overline{PQ}$ is a chord of the circle and $PQ = 6$ centimeters. If the shortest distance from $\overline{PQ}$ to O is 1 centimeter, what is the area of the circle in square centimeters?

(A) 5π

(B) 9π

(C) 10π

(D) 37π

(E) 100π

Begin by sketching out the information from the question like this:

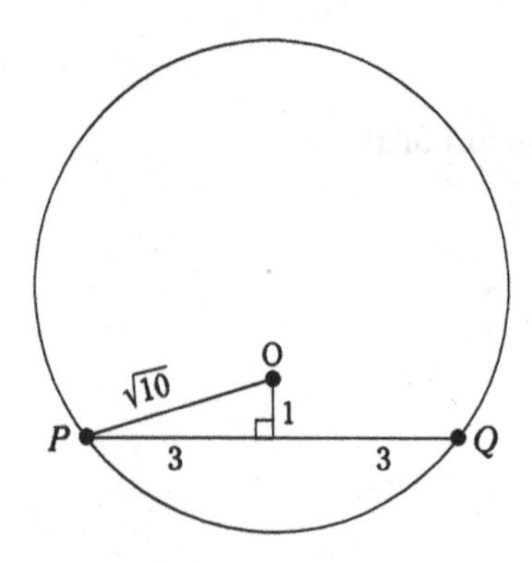

Notice that the line segment from O to $\overline{PQ}$ meets at a right angle. Furthermore, it bisects $\overline{PQ}$. After the radius is drawn ($\overline{OQ}$ or $\overline{OP}$), the result is a right triangle whose legs are 1 and 3 in length, so you can use the Pythagorean theorem to find the hypotenuse:

$$a^2 + b^2 = c^2$$
$$1^2 + 3^2 = c^2$$
$$1 + 9 = c^2$$
$$10 = c^2$$
$$\sqrt{10} = c$$

The hypotenuse of this triangle is also the radius of the circle, so use the formula for the area of a circle:

$$A = \pi r^2 = \pi\left(\sqrt{10}\right)^2 = 10\pi$$

The correct answer is Choice (C).

Examining Solid Geometry

Solid geometry is geometry in three dimensions (3D). It includes everything you know from plane geometry, but it also adds solids such as cubes, boxes, prisms, cylinders, pyramids, cones, and spheres.

Generally speaking, an ACT question probably won't require you to remember an obscure formula to solve a problem. However, the question may provide a useful formula and ask you to apply it to get an answer. So to feel completely comfortable applying them, make sure you familiarize yourself with a few formulas from solid geometry.

The most important formula for any solid is usually the formula for *volume* — the amount of space that the solid occupies. Volume is the 3D equivalent of area and is expressed in cubic units (such as cubic feet, cubic centimeters, and so forth).

For certain solids, however, you may need to work with their *surface area,* which is (as you may guess) the area of its total surface. You express surface area in square units (such as square inches, meters, and so forth).

In this section, I discuss everything you need to know about solid geometry to do well on the ACT.

Focusing on cubes and boxes

One of the most common solids you'll find popping up in ACT questions is the *cube,* which is a box that has six square faces. Both the volume and the surface area of a cube are based on the length of one side (s):

- **Volume of a cube:** $V = s^3$
- **Surface area of a cube:** $A = 6s^2$

EXAMPLE

The volume of a cube is k cubic meters, and its surface area is $\frac{3}{4}k$ square meters. What length is one side of the cube?

(A) 4 meters
(B) 6 meters
(C) 8 meters
(D) $2\sqrt{2}$ meters
(E) $4\sqrt{2}$ meters

The volume of this cube is k, so plug this value into the volume formula that I introduce earlier in the section:

$$k = s^3$$

The surface area of this cube is $\frac{3}{4}k$, so plug this value into the surface-area formula and solve for k in terms of s:

$$A = 6s^2$$
$$\frac{3}{4}k = 6s^2$$
$$3k = 24s^2$$
$$k = 8s^2$$

The right side of this last equation now equals k, so, using the previous equation, substitute s^3 for k and then solve for s.

$$s^3 = 8s^2$$
$$s = 8$$

So the side of this cube is 8 meters, making the correct answer Choice (C).

The volume and surface area of a box (rectangular solid) is based on its length (l), width (w), and height (h), as shown in these formulas:

- **Volume of a box:** $V = lwh$
- **Surface area of a box:** $A = 2lw + 2lh + 2wh$

A rectangular box whose dimensions are n, $10n$, and $12n$ has a volume of 15. What is the value of n?

(F) 2

(G) $\frac{1}{2}$

(H) $\sqrt{2}$

(J) $2\sqrt{2}$

(K) $\frac{\sqrt{2}}{2}$

Plug the volume and the length, width, and height (in any order) into the formula for the volume of a box and solve for n:

$$V = lwh$$
$$15 = (n)(10n)(12n)$$
$$15 = 120n^3$$
$$\frac{1}{8} = n^3$$

At this point, solve for n by taking the cube root of both sides:

$$\frac{1}{2} = n$$

The correct answer is Choice (G).

Incorporating spheres into your geometric repertoire

A *sphere* is the set of points in space that are all the same distance from a single point. A ball or globe is a perfect visual aid for a sphere. The formulas for both the volume and surface area of a sphere are based on its radius (r):

- **Volume of a sphere:** $V = \frac{4}{3}\pi r^3$
- **Surface area of a sphere:** $A = 4\pi r^2$

EXAMPLE

An artist is incorporating a globe into a large sculpture. They intend to paint the surface of the globe, which is a sphere with a volume of 36π cubic feet. What is the surface area of the globe?

(A) 9π

(B) 16π

(C) 36π

(D) 64π

(E) 81π

First, use the formula for the volume of the globe to find the radius:

$$V = \frac{4}{3}\pi r^3$$
$$36\pi = \frac{4}{3}\pi r^3$$
$$108\pi = 4\pi r^3$$
$$27 = r^3$$
$$3 = r$$

Now use the formula for the surface area of a sphere:

$$A = 4\pi r^2 = 4\pi(3)^2 = 36\pi$$

So the correct answer is Choice (C).

Figuring the volume of other solids (prisms, cylinders, pyramids, and cones)

Although cubes, boxes, and spheres (which I discuss in the previous section) are the most common geometric solids on the ACT, you may face a question with one of the less-common solids. In this section, I show you how to work with prisms, cylinders, pyramids, and cones. Because of their similar shapes, prisms and pyramids can be lumped together and cylinders and cones can be as well. So I discuss each of these solids in pairs in the following sections.

Prisms and pyramids

The volume of a prism and a pyramid is based on its height (h) and the area of its base (A_b), regardless of the shape of that base. Here are the formulas to use:

- **Volume of a prism:** $V = A_b h$
- **Volume of a pyramid:** $V = \frac{1}{3}A_b h$

To find the area of the base, you may need to use one of the other formulas for area from earlier in this chapter. For example, if the base of a solid is a triangle, use the formula for the area of a triangle: $A = \frac{1}{2}bh$.

EXAMPLE

The solid in the following figure is a prism with irregular pentagonal base. If the height of the prism is h units and its volume is $7h$ cubic units, what is the area of the base in square units?

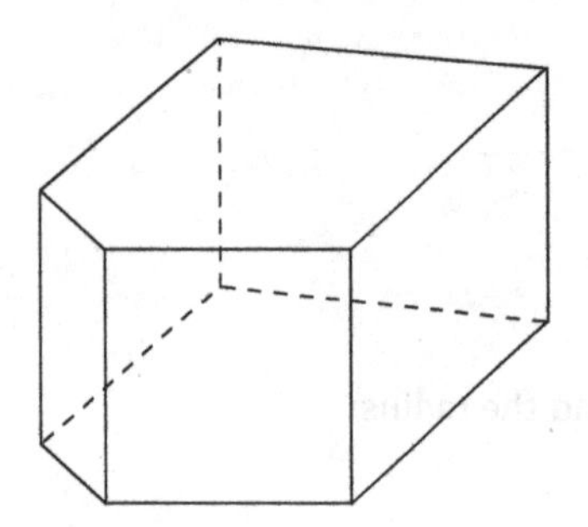

(A) 7

(B) $7h$

(C) $7h^2$

(D) $7h^3$

(E) Cannot be determined from the information given.

Plug the volume ($7h$) and height (h) into the formula for a prism and solve for A_b:

$$V = A_b h$$
$$7h = A_b h$$
$$7 = A_b$$

Thus, the correct answer is Choice (A).

Cylinders and cones

The volume of a cylinder and the volume of a cone are both based on the radius (r) of its circular base and its height (h). Check out the formulas:

- **Volume of a cylinder:** $V = \pi r^2 h$
- **Volume of a cone:** $V = \frac{1}{3}\pi r^2 h$

TIP

Notice that both of these formulas include πr^2, which is the area formula for a circle. So if you know the area of the base (A_b), you can substitute this value for πr^2 in the formula.

EXAMPLE

A cylinder and a cone both have the same volume and height. If the base of the cylinder has an area of 16π square inches, what is the radius of the cone?

(A) $\frac{3}{4}$

(B) $\frac{4}{3}$

(C) $\sqrt{3}$

(D) $2\sqrt{3}$

(E) $4\sqrt{3}$

The area of the cylinder's base is 16π, so plug this value into the formula for the volume of a cylinder:

$$V = \pi r^2 h$$
$$V = 16\pi h$$

The volumes of the cylinder and the cone are the same, so plug in $16\pi h$ for V into the equation for the volume of a cone and simplify:

$$V = \frac{1}{3}\pi r^2 h$$
$$16\pi h = \frac{1}{3}\pi r^2 h$$
$$16h = \frac{1}{3}r^2 h$$

The heights of the cylinder and the cone are equal, so you can cancel out h on both sides of the equation and solve for r:

$$16 = \frac{1}{3}r^2$$
$$48 = r^2$$
$$\sqrt{48} = r$$
$$4\sqrt{3} = r$$

So the correct answer is Choice (E).

IN THIS CHAPTER

- » Getting a refresher on trigonometry
- » Understanding basic matrix operations
- » Unraveling logarithms
- » Working with imaginary numbers

Chapter 12 Trig and Beyond: ACT Advanced Math Topics

The ACT includes some advanced math topics not found on the SAT. Don't fret, though. In this chapter, I get you up to speed. I introduce you to some basic concepts in trigonometry, including trig ratios, radian measure, sine and cosine graphs, and trig identities. I also show you how to apply basic math operations to matrices and how to calculate the determinant of a 2×2 matrix. Finally, I help you become comfortable working with both logarithms and imaginary and complex numbers.

TIP

I discuss these topics in order of their importance on the ACT: You'll most likely face three or four trigonometry questions, one matrix question, and one question testing you on either logarithms or imaginary numbers. So if you're pressed for time and need to choose among them, don't skip the trig!

Trigonometry: Watching the Sines and Taking a Few Tangents

Trigonometry — which literally means "the study of triangles" — builds on the basic geometry of triangles that you discover in Chapter 11. In that chapter, I discuss two important triangles: the 45-45-90 triangle and the 30-60-90 triangle, which are both shown in Figure 12-1.

REMEMBER

Any two triangles that have all congruent angles and sides in proportion are called *similar triangles.* The concept of similar triangles plays a vital role in trigonometry.

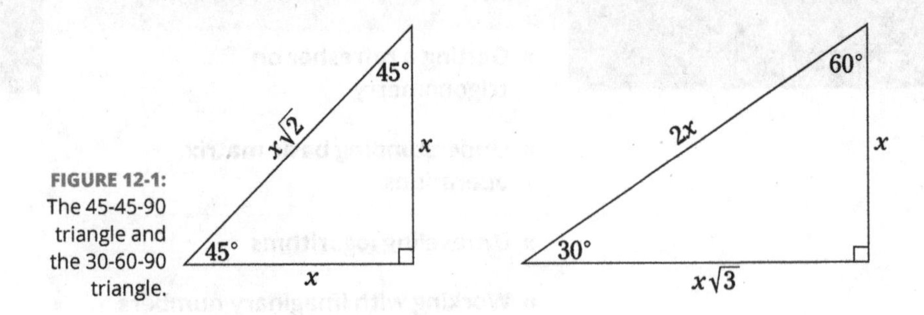

FIGURE 12-1: The 45-45-90 triangle and the 30-60-90 triangle.

For example, all 30-60-90 triangles have three sides that are in a ratio of $1:\sqrt{3}:2$. The additional fact that all 30-60-90 triangles are similar has an important implication: The ratio between any two corresponding sides of a 30-60-90 triangle is always the same, regardless of the size of the triangle. Figure 12-2 shows such triangles.

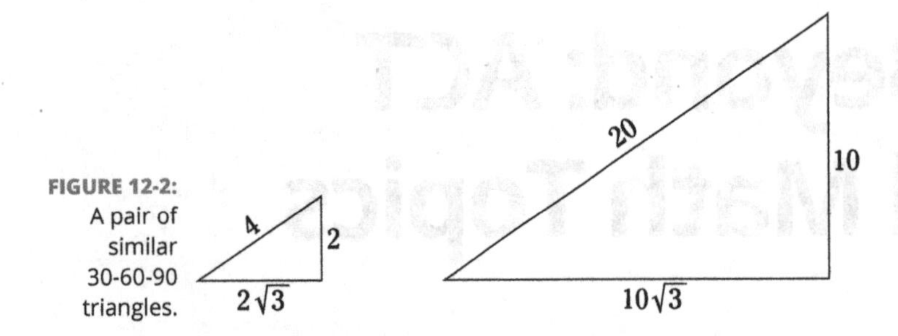

FIGURE 12-2: A pair of similar 30-60-90 triangles.

For each triangle in the figure, here's the ratio of the short leg to the hypotenuse:

$$\text{Smaller triangle: } \frac{\text{Short leg}}{\text{Hypotenuse}} = \frac{2}{4} = \frac{1}{2}$$

$$\text{Larger triangle: } \frac{\text{Short leg}}{\text{Hypotenuse}} = \frac{10}{20} = \frac{1}{2}$$

As you can see, the ratio of the short leg to the hypotenuse of a 30-60-90 triangle is always $\frac{1}{2}$, regardless of the size of the triangle. This result occurs because all 30-60-90 triangles are similar to each other.

Picking out the opposite, the adjacent, and the hypotenuse

What's great about similar triangles is that if you know the measure of one angle in any right triangle, you can find the *ratio* of any two sides even if you don't know the length of any of the sides. The first step is to name the three sides of a triangle in relation to the angle that you do know.

- The *hypotenuse* is the longest side, and it's always directly across from the right angle.
- The *opposite* is the leg of the triangle that isn't part of the angle.
- The *adjacent* is the leg of the triangle that is part of the angle.

In the triangle in Figure 12-3, you're working with angle *x*, which allows you to distinguish the two legs of the triangle in relation to this angle as the *opposite* (*O*) and the *adjacent* (*A*). The *hypotenuse* (*H*) of the triangle is always its longest side.

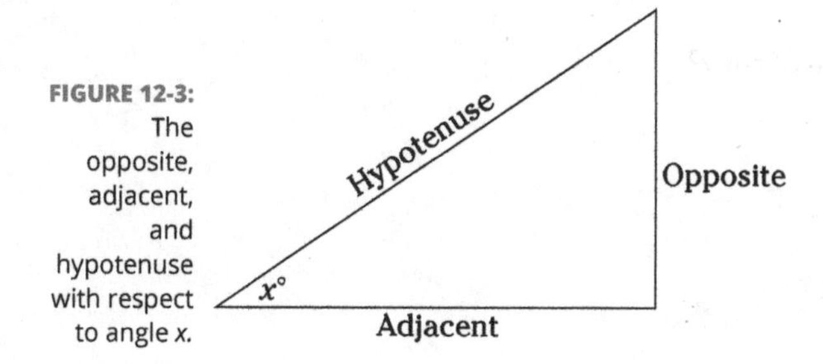

FIGURE 12-3: The opposite, adjacent, and hypotenuse with respect to angle *x*.

Knowing how to SOH CAH TOA

Every triangle has three distinct sides — the *opposite*, the *adjacent*, and the *hypotenuse* — giving rise to six distinct ratios among these sides. Table 12-1 lists these ratios and their formulas.

TABLE 12-1 The Six Trigonometric Ratios

Trig Ratios	Ratio of Sides	Formula
Sine (sin)	Opposite/Hypotenuse	$\sin x = \frac{O}{H}$
Cosine (cos)	Adjacent/Hypotenuse	$\cos x = \frac{A}{H}$
Tangent (tan)	Opposite/Adjacent	$\tan x = \frac{O}{A}$
Cotangent (cot)	Adjacent/Opposite	$\cot x = \frac{A}{O}$
Secant (sec)	Hypotenuse/Adjacent	$\sec x = \frac{H}{A}$
Cosecant (csc)	Hypotenuse/Opposite	$\csc x = \frac{H}{O}$

TIP

If you have trouble remembering the six trig ratios, the famous (or infamous) mnemonic SOH-CAH-TOA may help you. This mnemonic stands for:

- **S**ine is **O**pposite over **H**ypotenuese.
- **C**osine is **A**djacent over **H**ypotenuese.
- **T**an is **O**pposite over **A**djacent.

Knowing these three ratios can help you remember the other three, which are simply reciprocals of the first three:

- $\text{Sin } x\left(\frac{O}{H}\right)$ is the reciprocal of $\csc x\left(\frac{H}{O}\right)$.
- $\text{Cos } x\left(\frac{A}{H}\right)$ is the reciprocal of $\sec x\left(\frac{H}{A}\right)$.
- $\text{Tan } x\left(\frac{O}{A}\right)$ is the reciprocal of $\cot x\left(\frac{A}{O}\right)$.

EXAMPLE

In the following figure, what is the value of cot x?

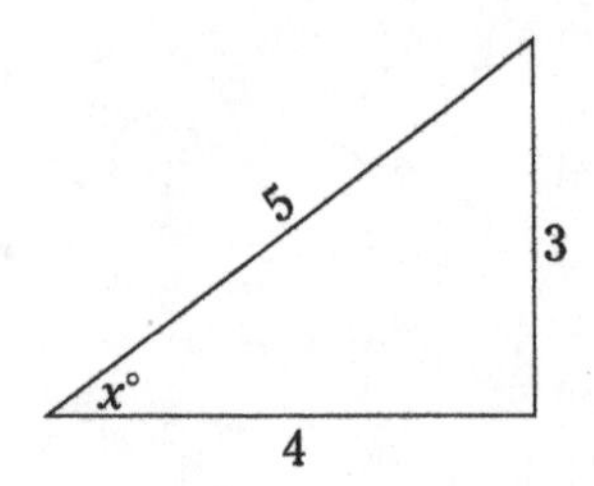

(A) $\frac{3}{4}$

(B) $\frac{3}{5}$

(C) $\frac{4}{5}$

(D) $\frac{4}{3}$

(E) $\frac{5}{4}$

The cotangent of an angle is the ratio of the adjacent side and the opposite side — that is:

$$\cot x = \frac{A}{O} = \frac{4}{3}$$

Thus, the correct answer is Choice (D).

EXAMPLE

In the following figure, which of the answer choices is equal to $\frac{5}{12}$?

EXAMPLE

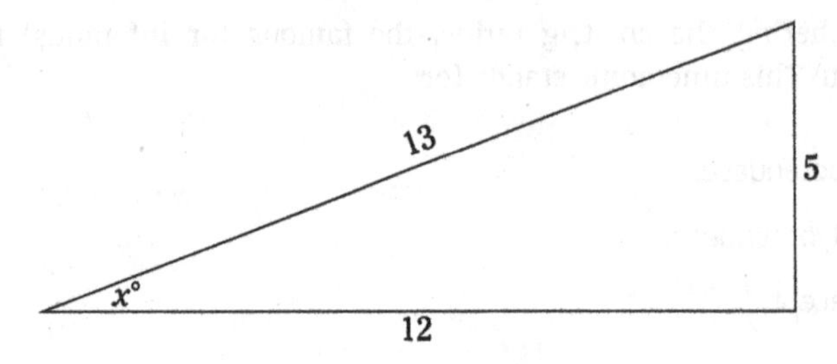

(F) $\sin x$

(G) $\cos x$

(H) $\tan x$

(J) $\cot x$

(K) $\sec x$

With respect to angle x in the triangle in the preceding figure:

$$\frac{5}{12} = \frac{O}{A} = \tan x$$

Therefore, the correct answer is Choice (H).

Feeling radiant with radian measure

Radian measure is an alternative way to measure angles. Instead of dividing the angle measurement of a circle into 360 degrees, a circle is divided into 2π radians. So

$$360° = 2\pi \text{radians}$$

REMEMBER

Unlike the symbol for degrees (°), the symbol for radians (*rad*) is rarely used in higher math: An angle without units of degrees is assumed to be expressed in radians. (Also, higher math uses radians almost exclusively.) Table 12-2 shows some conversions for the most commonly used angles.

TABLE 12-2 Radian Measure for Common Angles

Radians	Degrees
$\frac{\pi}{6}$	30°
$\frac{\pi}{4}$	45°
$\frac{\pi}{3}$	60°
$\frac{\pi}{2}$	90°
π	180°
2π	360°

You can use the following proportion for converting other angles from radians to degrees (or vice versa):

$$\frac{180}{\pi} = \frac{\text{degrees}}{\text{radians}}$$

EXAMPLE

What is the value of 75 degrees expressed as radians?

(A) $\frac{2\pi}{5}$

(B) $\frac{3\pi}{10}$

(C) $\frac{5\pi}{6}$

(D) $\frac{5\pi}{12}$

(E) $\frac{5\pi}{24}$

Use the formula, substituting 75 for *degrees* and r for *radians:*

$$\frac{180}{\pi} = \frac{75}{r}$$

Solve for *r:*

$$180r = 75\pi$$

$$r = \frac{75\pi}{180}$$

$$r = \frac{5\pi}{12}$$

So the correct answer is Choice (D).

Graphing trig functions

Although all six trig functions can be graphed, the ACT mostly requires you to work with graphs of sines and cosines. Both of these ratios have similar wave-shaped curves, as shown in Figure 12-4.

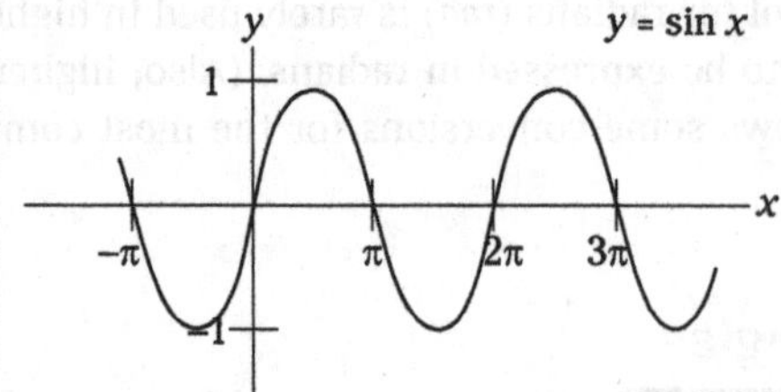

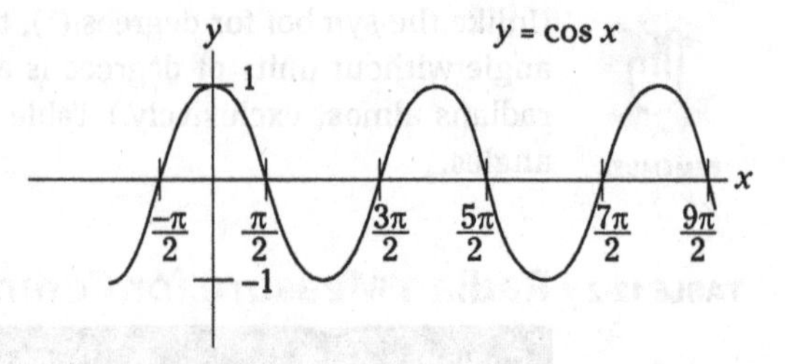

FIGURE 12-4: Graphs of $y = \sin x$ and $y = \cos x$, with periods and amplitudes labeled.

Radian measure (which I introduce in the earlier section, "Feeling radiant with radian measure") is typically used for graphing trig functions. Here are a few important words for describing waves, some of which may show up on your ACT:

- **Crest:** A *crest* (or *peak*) of the function is the highest point on the wave. For example, the function $\sin x$ in the figure has peaks at $\frac{\pi}{2}, \frac{5\pi}{2}, \frac{9\pi}{2}$, and so forth.
- **Trough:** A *trough* is the lowest point on the wave. For example, the function $\sin x$ has troughs at $\frac{3\pi}{2}, \frac{7\pi}{2}, \frac{11\pi}{2}$, and so on.
- **Period:** The *period* of the function is the horizontal distance between two adjacent crests. For example, the period of both $\sin x$ and $\cos x$ in the figure is 2π.
- **Amplitude:** The *amplitude* of the function is the vertical distance from the vertical midpoint of the function to the highest point on the function — that is, *half* the vertical distance from crest to trough. For example, the amplitude of both $\sin x$ and $\cos x$ is 1.

Use the following graph for the next two example questions:

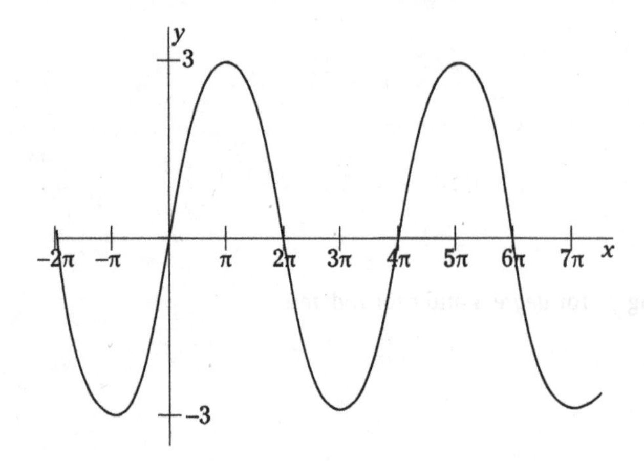

EXAMPLE

What is the period of this function?

(A) 2

(B) 3

(C) 6

(D) 2π

(E) 4π

The period is the distance from crest to crest, so this distance in the figure is 4π. Therefore, the correct answer is Choice (E).

EXAMPLE

What is the amplitude of this function?

(F) 2

(G) 3

(H) 6

(J) 2π

(K) 4π

The amplitude is *half* the vertical distance from crest to trough. This distance in the preceding figure is 3, making the correct answer Choice (G).

A more difficult problem may give you an equation and ask you to identify characteristics of the resulting graph.

EXAMPLE

Which of the following is the period of the function $y=2\cos x$?

(A) $\frac{\pi}{4}$

(B) $\frac{\pi}{2}$

(C) π

(D) 2π

(E) 4π

This is a transformation of the basic function $y = \cos x$, shown in Figure 12-4. You can use this basic function to sketch the graph you're looking for by plotting a few points. Either build a table on your calculator or just input a few point values for 2 sin x to create a table like this one:

x	0	π	2π	3π	4π
y	2	-2	2	-2	2

This should be enough points to sketch the following graph:

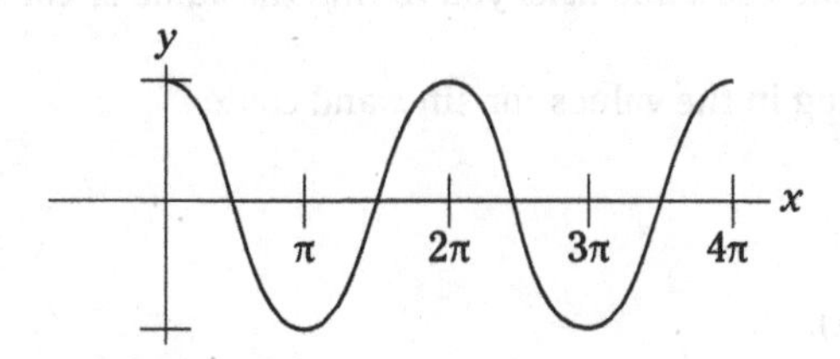

The distance between the two peaks is 2π, so the correct answer is Choice (D).

Identifying trig identities

Recall from the earlier section "Knowing how to SOH-CAH-TOA" that each of the six trig functions is a ratio of two of the three sides of a right triangle (the opposite, the adjacent, and the hypotenuse). These ratios create connections among the trig functions called *identities.* These identities are equations using the trig functions that are always true no matter what the value of the variable is. In a high school trig course, you may solve tricky problems that use more than a dozen trig identities. On the ACT, however, you'll probably need just five of them.

Here are the three *reciprocal identities* (so called because they show you that three pairs of trig function are reciprocals of each other):

$$\sin x = \frac{1}{\csc x} \qquad \cos x = \frac{1}{\sec x} \qquad \tan x = \frac{1}{\cot x}$$

To see why these identities are true, refer to the first identity in the preceding list and remember the SOH CAH TOA rule I introduce earlier in this chapter. This rule tells you that $\sin x = \frac{O}{H}$ and $\csc x = \frac{H}{O}$. So sin *x* and csc *x* are reciprocals of each other. You can make similar connections for the other two reciprocal identities.

The next two identities express the tangent and cotangent functions in terms of sines and cosines:

$$\tan x = \frac{\sin x}{\cos x} \qquad \cot x = \frac{\cos x}{\sin x}$$

Again, the SOH CAH TOA rule can help you make sense of why these are true. For example, here's how you use these rules to show why tan *x* is equal to $\frac{\sin x}{\cos x}$:

$$\frac{\sin x}{\cos x} = \frac{\frac{O}{H}}{\frac{A}{H}} = \frac{O}{A} = \tan x$$

The remaining identity follows easily, because tan *x* and cot *x* are reciprocals of each other, so cot *x* must equal $\frac{\cos x}{\sin x}$.

When you know these five trig identities, you can use them to answer ACT questions.

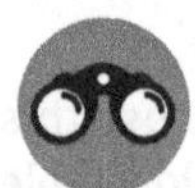

EXAMPLE

If $\sin x = 2a$ and $\cos x = 6b$, what is the value of cot *x*?

(A) $3ab$

(B) $12ab$

(C) $\frac{a}{b}$

(D) $\frac{a}{3b}$

(E) $\frac{3b}{a}$

The problem gives you the value of sin *x* and cos *x* and asks you to find the value of cot *x*.

Use the identity $\cot x = \frac{\cos x}{\sin x}$, and then plug in the values for sin *x* and cos *x*:

$$\cot x = \frac{\cos x}{\sin x} = \frac{6b}{2a} = \frac{3b}{a}$$

Therefore, the correct answer is Choice (E).

Following the laws: Laws of sines and cosines

The *law of sines* and the *law of cosines* allow you to measure the sides and angles of non-right triangles, such as the one shown in the following figure.

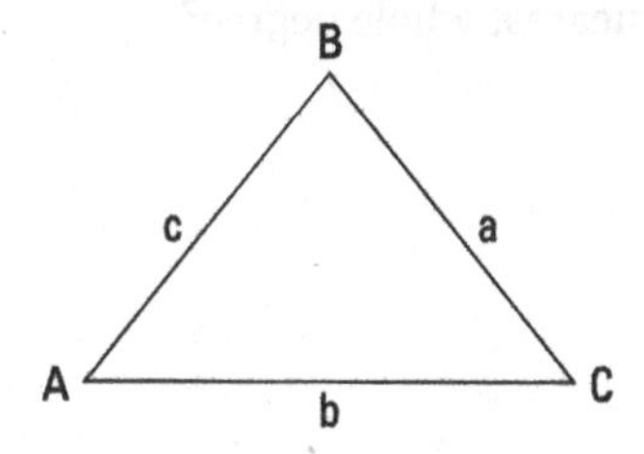

Notice that each angle A, B, and C is opposite its corresponding side a, b, and c. These six values are all used in the following two formulas:

Law of Sines :	***Law of Cosines :***
$\frac{\sin A}{a} = \frac{\sin B}{b} = \frac{\sin C}{c}$	$c^2 = a^2 + b^2 - 2ab\cos C$

When you know the lengths of all three sides of a triangle, the law of cosines allows you to find the measurement of the angles. For example:

Given that the law of cosines is $c^2 = a^2 + b^2 - 2ab\cos C$, what is the value of cos N in the following triangle?

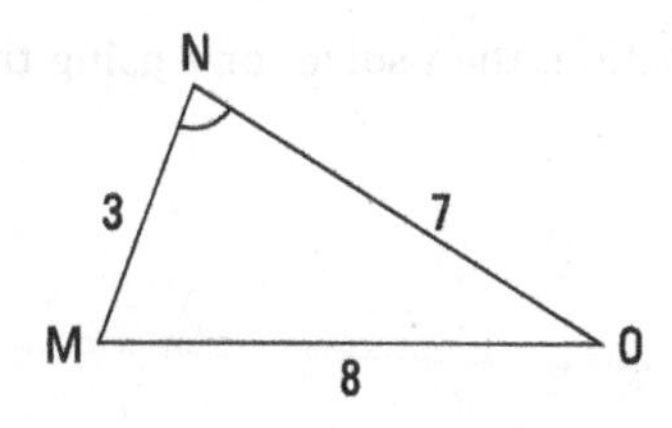

(A) $-\frac{1}{6}$

(B) $-\frac{1}{7}$

(C) $-\frac{1}{8}$

(D) $-\frac{1}{9}$

(E) $-\frac{1}{10}$

Use the law of cosines, making sure to choose N (the angle specified in the problem) as C, which makes C = 8. You can assign the values 3 and 7 to a and b in any order. Here's the result:

$$\begin{aligned} c^2 &= a^2 + b^2 - 2ab\cos C \\ 8^2 &= 3^2 + 7^2 - 2(3)(7)\cos N \\ 64 &= 58 - 42\cos N \\ 6 &= -42\cos N \\ -\frac{1}{7} &= \cos N \end{aligned}$$

Therefore, the correct answer is Choice (B).

In contrast, the law of sines is useful only when you know the measurement of at least one angle. For example:

In the following triangle, what is the value of $x°$, to the nearest whole degree?

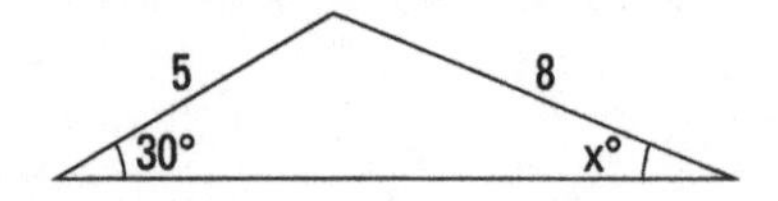

(F) 18°

(G) 19°

(H) 20°

(J) 21°

(K) 22°

Apply the law of sines, plugging in values for the two sides of the triangle and the angle that you know:

$$\frac{\sin A}{a} = \frac{\sin B}{b}$$

$$\frac{\sin 30°}{8} = \frac{\sin x°}{5}$$

$$\frac{5 \sin 30°}{8} = \sin x°$$

Now, use $\sin 30° = \frac{1}{2}$, to simplify the left side of the equation, then solve for x using the inverse sine function on your calculator:

$$\sin x° = 0.3125$$

$$\arcsin(\sin x°) = \arcsin 0.3125$$

$$x° \approx 18°$$

Therefore, the correct answer is Choice (F).

Reloading the Matrix

A *matrix* is a grid of numbers or variables arranged in horizontal rows and vertical columns. Figure 12-5 shows a variety of matrices.

$$\begin{bmatrix} 2 & 3 & 9 \\ -1 & 12 & -7 \end{bmatrix} \qquad \begin{bmatrix} -6x \\ -10x \\ 2x \end{bmatrix} \qquad \begin{bmatrix} 0 & 1 & 1 & 5 \\ 1 & 1 & 2 & -1 \\ 5 & 1 & 0 & 0 \\ 7 & -3 & 3 & 7 \end{bmatrix}$$

FIGURE 12-5: Examples of three matrices.

The *dimensions* of the matrices (that is, the number of rows and columns) in the figure are 2×3, 3×1, and 4×4, respectively. By convention, the number of rows is stated first and the number of columns second. (This convention becomes important when multiplying two matrices, as you discover later in this section.)

Matrices have a wide variety of uses in math beyond the scope of the ACT. For the test, you only need to know a few basics. I tell you exactly what to focus on in this section.

Adding and subtracting matrices

For the ACT, you definitely want to know how to add and subtract matrices. You're bound to encounter problems with these simple operations. Luckily, adding matrices is easy: Just add pairs of corresponding numbers in the matrix and place the results in a new matrix.

REMEMBER

The two matrices you're adding *must* have identical horizontal and vertical dimensions. As a result, the new matrix always has the same dimensions as the two matrices you're adding.

EXAMPLE

Which of the following is the result when you add $\begin{bmatrix} 5 & -6 \\ 2 & 0 \end{bmatrix}$ and $\begin{bmatrix} 3 & 4 \\ -3 & -1 \end{bmatrix}$?

(A) 4

(B) [4]

(C) $\begin{bmatrix} 8 & -2 \\ -1 & -1 \end{bmatrix}$

(D) $\begin{bmatrix} 3 & 4 & 5 & -6 \\ -3 & -1 & 2 & 0 \end{bmatrix}$

(E) Cannot be determined from the information given.

To find the correct answer, simply add each pair of corresponding numbers (3+5=8, 4+(−6)=−2, −3+2=−1, and −1+0=−1) and then place them in the same positions in a new matrix. The correct answer is Choice (C).

Subtracting matrices is similar to adding them: For each pair of corresponding numbers in the matrix, subtract the second number from the first and then place the results in a new matrix. As with addition, you can only subtract one matrix from another if both matrices have identical horizontal and vertical dimensions.

Which of the following is the result when you perform this matrix subtraction: $\begin{bmatrix} 7 & 1 \\ -2 & 0 \end{bmatrix} - \begin{bmatrix} 2 & -6 \\ 0 & -3 \end{bmatrix}$?

(F) $\begin{bmatrix} 5 & 7 \\ 2 & 3 \end{bmatrix}$

(G) $\begin{bmatrix} 5 & 7 \\ 2 & -3 \end{bmatrix}$

(H) $\begin{bmatrix} 5 & 7 \\ -2 & 3 \end{bmatrix}$

(J) $\begin{bmatrix} -5 & -7 \\ 2 & -3 \end{bmatrix}$

(K) $\begin{bmatrix} -5 & -7 \\ -2 & 3 \end{bmatrix}$

Subtract each pair of corresponding numbers in order (7−2=5, 1−(−6)=7, −2−0=−2, and 0−(−3)=3) and then place them in the same positions in a new matrix. The correct answer is Choice (H).

Making sense of matrix multiplication

You may encounter ACT questions that ask you to use multiplication with matrices. You may have to multiply an entire matrix by a number (very simple), or you may have to multiply two matrices by one another (not so simple). I show you how to do both in this section.

Multiplying a matrix by a real number (scalar multiplication)

Multiplying a matrix by a real number, also called *scalar multiplication*, is straightforward: Simply multiply the integer by every element in the matrix. Doing so is similar to using the distributive property to multiply a number by each term inside a set of parentheses. Yeah, it's that easy.

Which of the following is the result when you multiply the matrix $[1 \quad -2 \quad x]$ by 3?

(A) $-3+3x$

(B) $[-3 \quad 3x]$

(C) $[3 \quad -6 \quad x]$

(D) $[3 \quad -6 \quad 3x]$

(E) Cannot be determined from the information given.

To find the answer, you simply multiply each element of the matrix by 3. Place these three answers into a new matrix of the same dimensions:

$$3[1-2 \quad x]=[3-6 \quad 3x]$$

Therefore, the correct answer is Choice (D).

Multiplying two matrices

Multiplying two matrices by one another can be tricky, especially when the matrices themselves are lengthy. However, on the ACT, you'll probably just have to multiply pairs of matrices that have either one row or one column.

Matrix multiplication is *not* commutative, because the order in which you multiply two matrices can change the result. In other words, if P and Q are matrices, $P \times Q$ doesn't necessarily equal $Q \times P$.

In this section, I show you first how to multiply a matrix that has one column by a matrix that has one row. Then, I show you how to reverse this process and multiply a matrix that has one row by a matrix that has one column.

Multiplying a vertical matrix by a horizontal matrix

An easy way to multiply a vertical matrix by a horizontal matrix is to set up a small grid that resembles a multiplication table. This method allows you to fill in the numbers to get the right answer.

EXAMPLE

If $T=\begin{bmatrix}1\\5\\3\end{bmatrix}$ and $U=[2\ \ 0\ \ -4]$, which of the following is the result when you multiply $T \times U$?

(A) $[-10]$

(B) $[2\ \ 0\ \ -12]$

(C) $\begin{bmatrix}2\\0\\-12\end{bmatrix}$

(D) $\begin{bmatrix}2 & 0 & -4\\10 & 0 & -20\\6 & 0 & -12\end{bmatrix}$

(E) $\begin{bmatrix}2 & 10 & 6\\0 & 0 & 0\\-4 & -20 & -12\end{bmatrix}$

Note that *T* has one column and *U* has one row, so you can multiply these two matrices in this order. So set up a grid with the first matrix in the left column and the second matrix in the top row:

	2	0	–4
1			
5			
3			

For each box in the grid, multiply the two numbers in the corresponding row and column, like this:

	2	0	–4
1	2	0	–4
5	10	0	–20
3	6	0	–12

The contents of the grid is the result of the matrix multiplication:

$$\begin{bmatrix}2 & 0 & -4\\10 & 0 & -20\\6 & 0 & -12\end{bmatrix}$$

So the correct answer is Choice (D).

Multiplying a horizontal matrix by a vertical matrix

The process for multiplying a horizontal matrix by a vertical matrix is different from multiplying vertical ones by horizontal ones. This fact illustrates the idea that matrix multiplication is *not* commutative — when you reverse the order of the multiplication, the result is usually a different matrix. In the following example, you multiply the two matrices from the previous section in reverse order to see how this happens.

$$T=\begin{bmatrix}1\\5\\3\end{bmatrix} \qquad U=[2\ \ 0\ \ -4]$$

EXAMPLE

Which of the following is the result when you multiply $U \times T$?

(A) $[-10]$

(B) $[2 \ \ 0 \ \ -12]$

(C) $\begin{bmatrix} 2 \\ 0 \\ -12 \end{bmatrix}$

(D) $\begin{bmatrix} 2 & 0 & -4 \\ 10 & 0 & -20 \\ 6 & 0 & -12 \end{bmatrix}$

(E) $\begin{bmatrix} 2 & 10 & 6 \\ 0 & 0 & 0 \\ -4 & -20 & -12 \end{bmatrix}$

Notice that U has three columns and T has three rows, so you can multiply these two matrices in this order. However, the result will be different. To start, make another grid, this time with U in the left column and T in the top row:

$$\begin{array}{cc} & \begin{bmatrix} 1 \\ 5 \\ 3 \end{bmatrix} \\ \begin{bmatrix} 2 & 0 & -4 \end{bmatrix} & \boxed{} \end{array}$$

This time, the resulting grid has only one box. To fill it in, multiply the three pairs of corresponding numbers in the two matrices, and then add the results:

$$(2\times1)+(0\times5)+(-4\times3)=2+0+(-12)=-10$$

Place this result into the box:

$$\begin{array}{cc} & \begin{bmatrix} 1 \\ 5 \\ 3 \end{bmatrix} \\ \begin{bmatrix} 2 & 0 & -4 \end{bmatrix} & \boxed{-10} \end{array}$$

The contents of the grid become the result of the matrix multiplication:

$$[-10]$$

So the correct answer is Choice (A).

Determining the meaning of determinants

A *determinant* is a common operation performed on a square matrix. On the ACT, the only determinant formula you need to be familiar with is for a 2×2 matrix. Here's the formula for the determinant of $M=\begin{bmatrix} a & b \\ c & d \end{bmatrix}$:

$$\text{Det}(M) = ad-bc$$

Note that the determinant of a matrix is simply a number, not a matrix. To solve a problem with a determinant, you simply plug the numbers from the matrix into the formula and solve.

What is the determinant of the matrix $\begin{bmatrix} 4 & 5 \\ 6 & 7 \end{bmatrix}$?

(A) −22

(B) −2

(C) 0

(D) 2

(E) 22

First, apply the determinant formula:

$$ad-bc=(4)(7)-(5)(6)$$

Now evaluate to solve the problem:

$$=28-30=-2$$

Therefore, the correct answer is Choice (B).

What is the value of k if the determinant of the matrix $\begin{bmatrix} k & 2k \\ k^2 & 0 \end{bmatrix}$ equals 16?

(F) −2

(G) −1

(H) 0

(J) 1

(K) 2

Start by applying the determinant formula and then evaluate:

$$ad-bc=(k)(0)-(2k)(k^2)=0-2k^3=-2k^3$$

This value equals 16, so create your equation and solve:

$$\begin{aligned} -2k^3 &= 16 \\ k^3 &= -8 \\ k &= -2 \end{aligned}$$

So the correct answer is Choice (F).

Logging Some Hours with Logarithms

A *logarithm* (*log* for short) is an alternative way of working with powers (exponents). The logs you're likely to see on the ACT won't be as difficult as those you may have worked with in your math classes. Still, logs can be a little confusing. To help you out, in this section, I give you a useful way to think about logarithms and then show you how to apply this idea.

Understanding logarithms

Like a root (or radical), a logarithm is an inverse operation because it will undo a power. Recall from Chapter 4 that a power has two parts: a *base* (bottom number) and an *exponent* (top number). A root will undo the exponent and give you back the base you started with. For example:

$5^3=125 \qquad \sqrt[3]{125}=5$

In this example, I started with a base of 5 and an exponent of 3. The power operates on these two numbers, giving you 125. The root "undoes" this operation, eliminating the exponent of 3 and giving you back the base you started with — that is, 5. (I know "undoes" isn't exactly great word choice, but work with me here.)

In contrast, a logarithm will undo the *base* and give you back the *exponent* you started with:

$5^3=125 \qquad \log_5 125=3$

This time I started with the same base of 5 and exponent of 3 but used a logarithm to reverse the operation in a different way, eliminating the base of 5 and giving you back the exponent you started with — that is, 3.

REMEMBER

On the ACT, your main task with logarithms is to remove them as quickly as possible. The best way to do that is to turn any equation that has a messy logarithm into one that has a nice, clean power. When you do this, you'll usually find that the problem looks a lot easier, and you can take it from there without much trouble.

TIP

A good way to avoid the confusion of logarithms is to memorize this fact:

$\log_{10}100=2 \qquad$ means $\qquad 10^2=100$

When I solve a logarithm problem, I always write this note down so I can refer to it. Believe me, it helps reduce the confusion.

EXAMPLE

What is the value of n if $\log_4 n=3$?

(A) 64

(B) 81

(C) $\sqrt[3]{4}$

(D) $\sqrt[4]{3}$

(E) Cannot be determined from the information given.

The only thing really difficult about this problem is the fact that it includes a logarithm. So your best bet is to get rid of it. To do this, use the tip I just provided to turn this log equation into a power equation:

$\log_4 n=3 \qquad$ means $\qquad 4^3=n$

Now the problem practically solves itself:

$n=4^3=64$

Therefore, the correct answer is Choice (A).

Sometimes, you may have a little more work ahead of you, even after you remove the log. But in almost every case, changing the log to a power turns the problem into something you know how to handle.

EXAMPLE

If $\log_p 4 = 2$, what is the value of p^4?

(F) 4

(G) 16

(H) 256

(J) $\frac{1}{4}$

(K) $\frac{1}{16}$

Turn the log equation into a power equation:

$$\log_p 4 = 2 \qquad \text{means} \qquad p^2 = 4$$

The equivalent equation $p^2 = 4$ is much more intuitive to work with, and you can solve it easily:

$$p^2 = 4$$
$$p = \pm 2$$

Thus, p is either 2 or –2. In either case, $p^4 = 16$, so the correct answer is Choice (G).

Identifying the common log of base 10

Whenever you see the symbol *log* without a subscript, you can assume that the base is 10. This notation is called the *common log*. For example:

EXAMPLE

What is the value of $\log 0.01$?

(A) -2

(B) -1

(C) $\frac{1}{2}$

(D) 1

(E) 2

TIP

Sometimes, writing the value base of 10 into a common log can help you clarify a problem:

$$\log 0.01 = \log_{10} 0.01$$

Now, set the value to x and rewrite the problem as an exponent:

$$\log_{10} 0.01 = x$$
$$10^x = 0.01$$

Rewriting the problem in this way helps you draw on your knowledge of exponents (see Chapter 4) to solve it:

$$x = -2 \qquad \text{because} \qquad 10^{-2} = 0.01$$

Therefore, the correct answer is Choice (A).

Following the laws of logarithms

The *laws of logarithms* (*laws of logs* for short) are a list of identities that can help you simplify and evaluate logarithmic expressions:

Law of Logarithms	*Example*
$\log_a a = 1$	$\log_{10} 10 = 1$
$\log_a 1 = 0$	$\log_2 1 = 0$
$\log_a \frac{1}{a} = -1$	$\log_5 \frac{1}{5} = -1$
$\log_a a^b = b$	$\log_3 3^4 = 4$
$\log_n a + \log_n b = \log_n ab$	$\log_{10} 1{,}000 + \log_{10} 100 = \log_{10} 100{,}000 = 5$
$\log_n a - \log_n b = \log_n \frac{a}{b}$	$\log_{10} 1{,}000 - \log_{10} 100 = \log_{10} \frac{1{,}000}{100} = 1$
$b \log_n a = \log_n a^b$	$9 \log 2 = \log 2^9$

EXAMPLE

What is the value of $\log_9 9 + \log_3 \frac{1}{3} + \log_3 9$?

(A) -2

(B) -1

(C) $\frac{1}{2}$

(D) 1

(E) 2

To begin, evaluate the three logs using the laws of logs: $\log_9 9 = 1$, $\log_3 \frac{1}{3} = -1$, $\log_3 9 = \log_3 3^2 = 2$

Substitute and solve:

$$\log_9 9 + \log_3 \frac{1}{3} + \log_3 9 = 1 - 1 + 2 = 2$$

Therefore, the correct answer is Choice (E).

A more difficult problem may require you to use the laws of logs for addition, subtraction, and exponents:

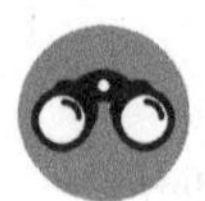

EXAMPLE

What is the value of $3\log_6 2 + 2\log_6 3 - \log_6 12$?

(F) -2

(G) -1

(H) $\frac{1}{2}$

(J) 1

(K) 2

To begin, use $b\log_n a = \log_n a^b$ to rewrite the three terms in the expression. I do this in two steps to make sense of the process:

$$3\log_6 2 + 2\log_6 3 - \log_6 12$$
$$= \log_6 2^3 + \log_6 3^2 - \log_6 12$$
$$= \log_6 8 + \log_6 9 - \log_6 12$$

Next, condense the first two terms down to one term using $\log_n a + \log_n b = \log_n ab$:

$$= \log_6 72 - \log_6 12$$

Now, simplify this expression using $\log_n a - \log_n b = \log_n \frac{a}{b}$, and complete the problem:

$$= \log_6 \frac{72}{12} = \log_6 6 = 1$$

Therefore, the correct answer is Choice (J).

Imagining the Possibilities with Imaginary Numbers

When you take the square root of a negative number, the result is an *imaginary number.* Imaginary numbers are ruled out of many math problems for a variety of reasons. On the ACT, however, you may face a question or two (though probably not more than that) that require you to understand imaginary numbers. In this section, I introduce you to why imaginary numbers exist in the first place, and then I show you a variety of ways to work with them.

Imaging how imaginary numbers arise

The creation of imaginary numbers begins with the leap of faith that they exist in the first place. The variable i is made to equal the square root of –1:

$$i = \sqrt{-1}$$

Note why this equation is a leap of faith: It says that when you multiply i by itself, the result is –1. But when you multiply any real number by itself, the result is never less than 0. So i isn't a real number.

Despite the uncomfortable fact that i isn't a real number, you can perform many operations on it as you would with a variable. For example, you can add i to itself:

$$i + i = 2i \qquad i + i + i = 3i \qquad i + i + i + i = 4i$$

Because you can add i to itself, you can multiply i by a constant using the normal rules of multiplication:

$$2 \times i = 2i \quad -5 \times i = -5i \qquad 0.1 \times i = 0.1i$$

You also can apply an exponent to i:

$i^2 = -1$ by definition

$i^3 = -i$ because $i^3 = i^2 \times i = -1 \times i = -i$

$i^4 = 1$ because $i^4 = i^2 \times i^2 = -1 \times (-1) = 1$

The identity $i^4 = 1$ implies that every exponent i that's divisible by 4 also equals 1:

$$i^8 = i^4 \cdot i^4 = 1 \qquad i^{12} = i^4 \cdot i^4 \cdot i^4 = 1 \qquad i^{16} = i^4 \cdot i^4 \cdot i^4 \cdot i^4 = 1$$

This insight allows you to calculate greater exponents of i with relative ease, as the next example illustrates:

EXAMPLE

What is the value of i^{234}?

(A) 1

(B) -1

(C) i

(D) $-i$

(E) Cannot be determined from the information given.

To begin, break i^{234} into two factors as follows:

$$i^{234} = i^{232} \cdot i^2$$

Now, notice that $i^{232} = 1$, because 232 is divisible by 4:

$$= 1 \cdot i^2$$

From here, answer the question using your knowledge of exponents of i:

$$= i^2 = -1$$

Therefore, the correct answer is Choice (B).

Working with complex numbers

When you add or subtract real and imaginary values, the result is a *complex number* — that is, a two-term expression of the form $a + bi$. For example:

Complex Number	*a Value*	*b Value*
$i+1$	1	1
$2-8i$	2	-8
$-6+3i$	-6	3
$-11-i$	-11	-1

REMEMBER

Working with complex numbers is similar to working with algebraic expressions (turn to Chapter 5 for more details).

For example, you can add complex numbers just as you would add binomials, by adding the coefficients of like terms:

$$(5 + 2i) + (-7 + 6i) = -2 + 8i$$

Similarly, you can subtract complex numbers by distributing the minus sign, and then combining like terms:

$$(5+2i)-(-7+6i)=5+2i+7-6i=12-4i$$

Here's an example that requires you to add and subtract complex numbers:

EXAMPLE

Which of the following is equivalent to $\frac{(3-5i)+(9-7i)}{(-2+3i)-(4-3i)}$?

(A) 2

(B) −2

(C) $1+i$

(D) $1-i$

(E) $2-2i$

To begin, simplify the numerator by adding and the denominator by subtracting:

$$\frac{(3-5i)+(9-7i)}{(-2+3i)-(4-3i)}=\frac{12-12i}{-6+6i}$$

Factor out 6 from both the numerator and denominator, and then cancel out this factor:

$$=\frac{12(1-i)}{-6(1-i)}=\frac{12}{-6}=-2$$

Therefore, the correct answer is Choice (B).

To multiply a pair of complex numbers, use the rules for FOILing that you know from Chapter 5, and then simplify using $i^2=-1$. The following example shows you how this process works:

EXAMPLE

Given that $i^2=-1$, what is the result when you multiply $4+3i$ by $-6+2i$?

(F) $-6+10i$

(G) $24+10i$

(H) $30+10i$

(J) $-24-10i$

(K) $-30-10i$

Multiply these two expressions using the FOIL rule (see Chapter 5):

$$(4+3i)(-6+2i)=-24+8i-18i+6i^2$$

Combine the two terms containing i ($-18i$ and $8i$):

$$=-24-10i+6i^2$$

Remember that $i^2=-1$, so you can simplify the i^2 term:

$$=-24-10i-6$$

And, finally, combine the two constant terms (−6 and −24):

$$=-30-10i$$

Thus, the correct answer is Choice (K).

In complex numbers, where $i^2 = -1$, which of the following is equivalent to $\frac{1}{1+i}+\frac{1}{1-i}$?

EXAMPLE

(F) 1

(G) 2

(H) i

(J) $\frac{1}{i}$

(K) $\frac{2}{i}$

Add these two fractions just as you would add two rational expressions in algebra. First, multiply the denominators to find the common denominator of $(1+i)(1-i)$. Then increase each fraction to adjust the denominators as needed:

$$\frac{1}{1+i}+\frac{1}{1-i}=\frac{1-i}{(1+i)(1-i)}+\frac{1+i}{(1+i)(1-i)}$$

Now add the fractions according to the usual rules: Add the numerators and keep the denominator the same. Your equation now looks like this:

$$=\frac{1-i+1+i}{(1+i)(1-i)}$$

Simplify the numerator by combining like terms:

$$=\frac{2}{(1+i)(1-i)}$$

Next, simplify the denominator by FOILing and combining like terms:

$$=\frac{2}{1+i-i-i^2}$$

$$=\frac{2}{1-i^2}$$

Remember that $i^2 = -1$, so now you can simplify even further:

$$=\frac{2}{1-(-1)}=\frac{2}{2}=1$$

So the correct answer is Choice (F).

Chapter 13
Practice Problems for Geometry, Trig, and Advanced Math

Are you interested in practicing some geometry and trigonometry questions from the material in Chapters 11 and 12? Well, look no further. In this chapter, I include 36 problems with completely worked-out answers at the end of the chapter. Get your pencil ready!

Geometry, Trig, and Advanced Math Practice Problems

1. In the following figure, the two horizontal lines are parallel, and $a° = 2b°$. Which of the following answer choices is true?

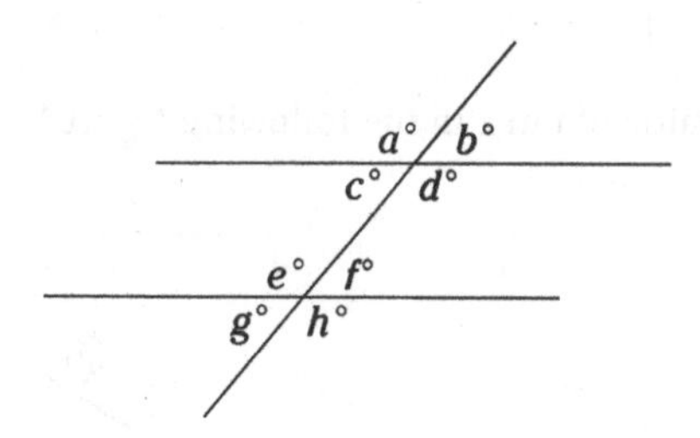

(A) $a° + d° = 200°$

(B) $a° + f° = 200°$

(C) $a° + e° = 240°$

(D) $b° + g° = 180°$

(E) $b° + h° = 120°$

2. In the following figure, if the area of ΔTUV is 34, what is the area of ΔTUW?

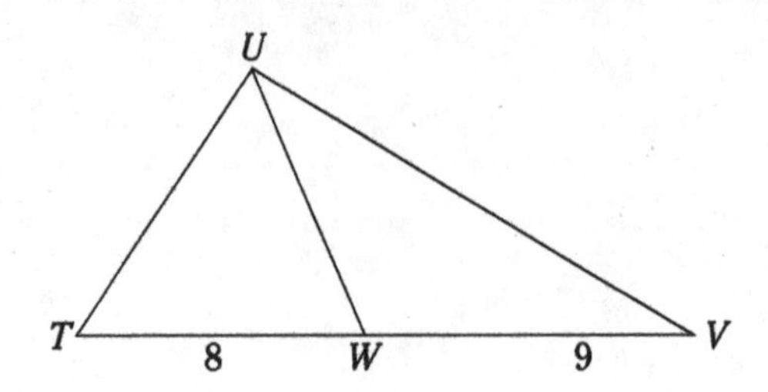

(F) 8

(G) 12

(H) 16

(J) 18

(K) 32

3. In the following figure, a circle with an area of 11π is embedded in a square. What is the perimeter of the square?

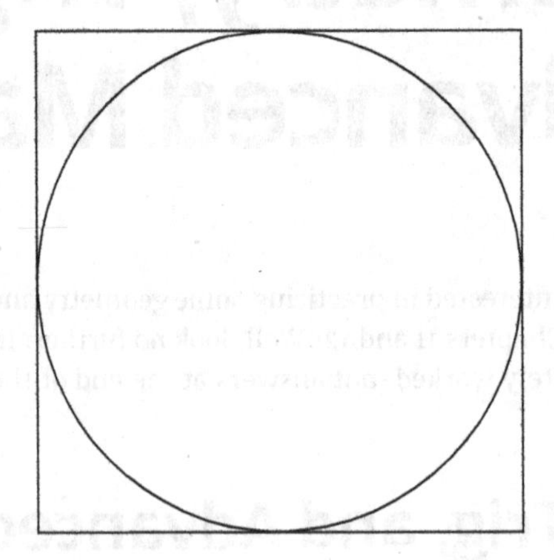

(A) 44

(B) 88

(C) $2\sqrt{11}$

(D) $4\sqrt{11}$

(E) $8\sqrt{11}$

4. What is the value of tan x in the following figure?

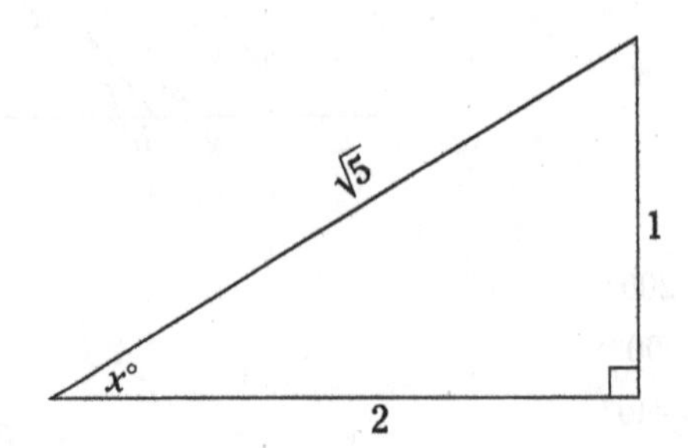

(F) 1

(G) 2

(H) $\sqrt{5}$

(J) $\frac{1}{2}$

(K) $\frac{\sqrt{5}}{5}$

5. What is the area of the trapezoid in the following figure?

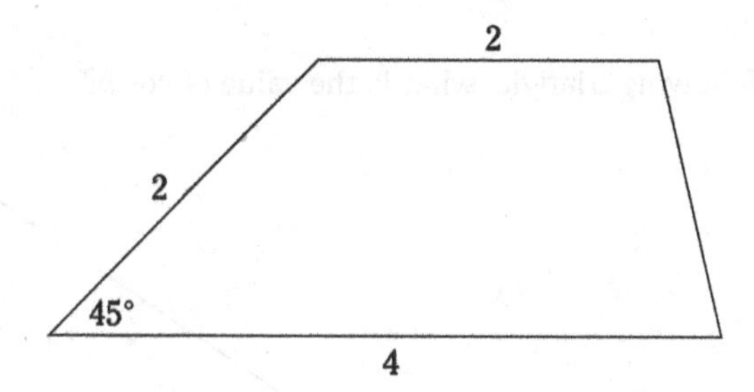

(A) 3

(B) 4

(C) 8

(D) $3\sqrt{2}$

(E) $6\sqrt{2}$

6. Kailani rides her bicycle to school and to her job at the mall. Her school is located exactly 4 miles due north of the mall and exactly 6 miles due east of her house. To the nearest whole mile, what is the shortest direct distance from Kailani's house to the mall?

(F) 6 miles

(G) 7 miles

(H) 8 miles

(J) 9 miles

(K) 10 miles

7. What is the result when the sum of the matrices $[3 \quad 4 \quad 1]$ and $[5 \quad -7 \quad 0]$ is multiplied by 4?

(A) 24

(B) $[24]$

(C) $[8 \quad -3 \quad 1]$

(D) $[19 \quad 9 \quad 4]$

(E) $[32 \quad -12 \quad 4]$

8. The four interior angles in a quadrilateral measure 70°, f°, $4f$°, and $(2f - 25)$°. What is the value of f?

(F) 25

(G) 35

(H) 40

(J) 45

(K) 50

9. A room has a square floor whose area is 625 square feet. If the volume of the entire room is 7,500 cubic feet, what is the difference between the length of one side of the room and the height of the room?

(A) 10 feet

(B) 11 feet

(C) 12 feet

(D) 13 feet

(E) 14 feet

10. In the following triangle, what is the value of cos b?

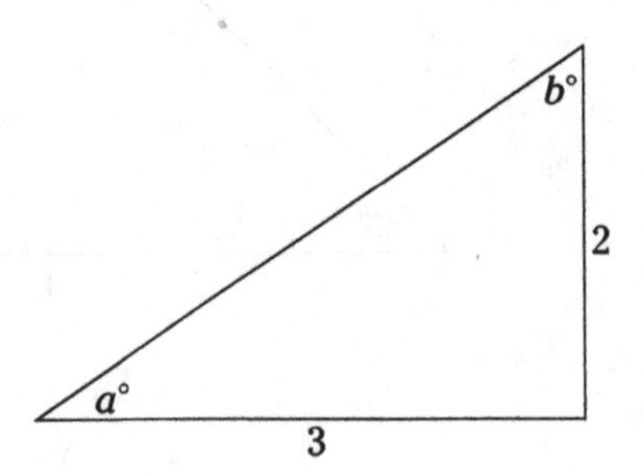

(F) $\sqrt{13}$

(G) $\frac{\sqrt{13}}{2}$

(H) $\frac{\sqrt{13}}{3}$

(J) $\frac{2\sqrt{13}}{13}$

(K) $\frac{3\sqrt{13}}{13}$

11. If the area of the parallelogram in the following figure is A, which of the answer choices is true?

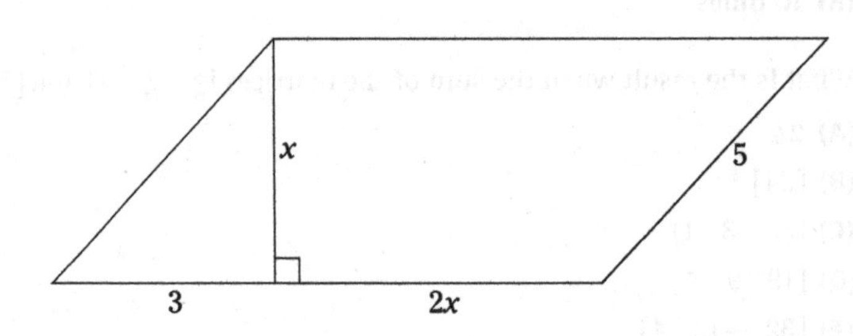

(A) $0 < A < 20$

(B) $10 < A < 30$

(C) $20 < A < 40$

(D) $30 < A < 40$

(E) $40 < A < 50$

12. A circular fountain with a circumference of 24 feet is surrounded by a 3-foot-wide cement walkway. What is the total distance in feet around the outer edge of this walkway?

(F) 27

(G) 30

(H) $12 + 3\pi$

(J) $24 + 3\pi$

(K) $24 + 6\pi$

13. What is the value of v^2 if $\log_3 v = 2$?

(A) 3

(B) 8

(C) 9

(D) 64

(E) 81

14. What is the length of a 40° arc of a circle with a diameter of 18?

(F) π

(G) 2π

(H) 3π

(J) 9π

(K) 10π

15. In the following figure, what is the value of x?

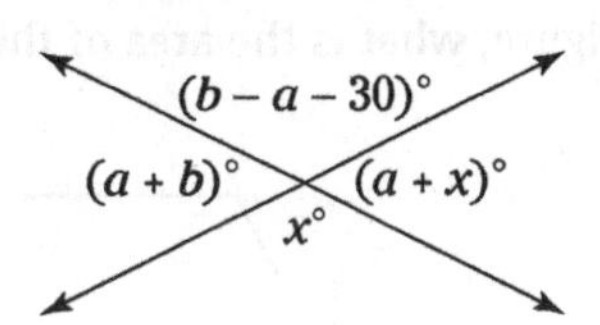

(A) 60

(B) 75

(C) 90

(D) 100

(E) 105

16. The formula for the surface area A of a sphere with a radius r is $A = 4\pi r^2$. What is the circumference of a circle that has the same radius as a sphere with a surface area of 20 square meters?

(F) $\pi\sqrt{5}$

(G) $2\pi\sqrt{5}$

(H) $\sqrt{5\pi}$

(J) $\sqrt{10\pi}$

(K) $2\sqrt{5\pi}$

17. If $\sin \phi = x$ and $\sec \phi = y$, what is the value of $\tan \phi$?

(A) xy

(B) $\frac{1}{x}$

(C) $\frac{1}{y}$

(D) $\frac{1}{xy}$

(E) Cannot be determined from the information given.

18. What is the area of an isosceles triangle with two sides of length 13 and one side of length 10?

(F) 60

(G) 65

(H) 100

(J) 120

(K) 130

19. If $i^2 = -1$, what is the result when you multiply $(1-i)$ by $(3+2i)$?

(A) $2+i$

(B) $3+i$

(C) $3-i$

(D) $4+i$

(E) $5-i$

20. In the following figure, what is the area of the parallelogram?

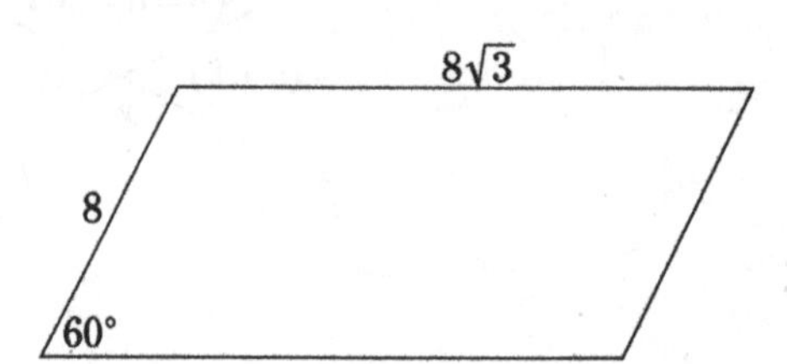

(F) 48

(G) 96

(H) $48\sqrt{3}$

(J) $64\sqrt{3}$

(K) $96\sqrt{3}$

21. If an angle measures $\frac{4\pi}{5}$ radians, what is its measurement in degrees?

(A) 67.5°

(B) 72°

(C) 135°

(D) 144°

(E) 288°

22. In the following figure, $\overline{PQ}$ is tangent to the circle and has a length of 15. If the circle has an area of 64π, what is the length of $\overline{OP}$?

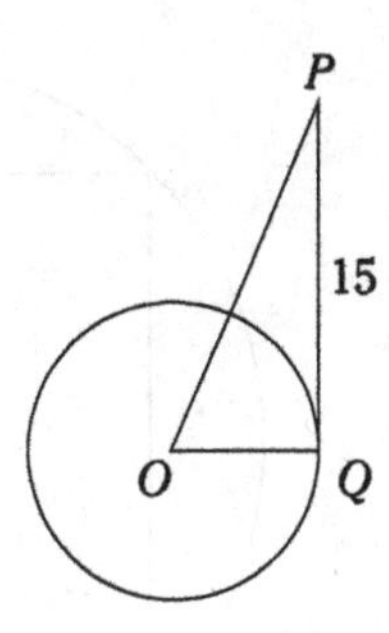

(F) 16

(G) 17

(H) 18

(J) 19

(K) 20

23. In the following figure, $p° =$

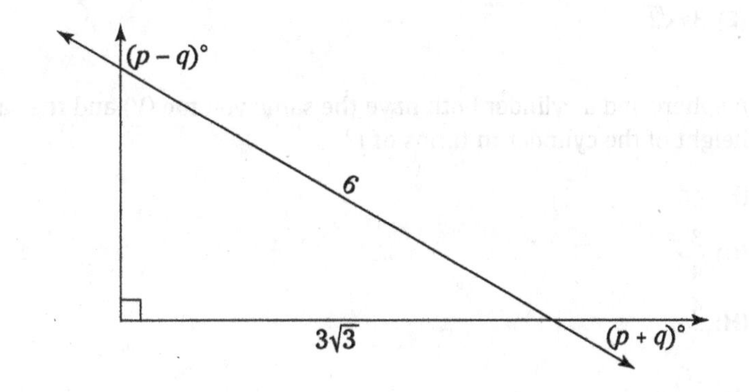

(A) 45°

(B) 55°

(C) 65°

(D) 75°

(E) 80°

24. If $A = \begin{bmatrix} 6 \\ 7 \end{bmatrix}$ and $B = \begin{bmatrix} -1 & 4 \end{bmatrix}$, what is the result of the matrix multiplication $A \times B$?

(F) 22

(G) $[22]$

(H) $\begin{bmatrix} -6 & 28 \end{bmatrix}$

(J) $\begin{bmatrix} -6 & 24 \\ -7 & 28 \end{bmatrix}$

(K) $\begin{bmatrix} -6 & -7 \\ 24 & 28 \end{bmatrix}$

25. In the following figure, the circle is circumscribed about the square. If the perimeter of the square is 12, what is the circumference of the circle?

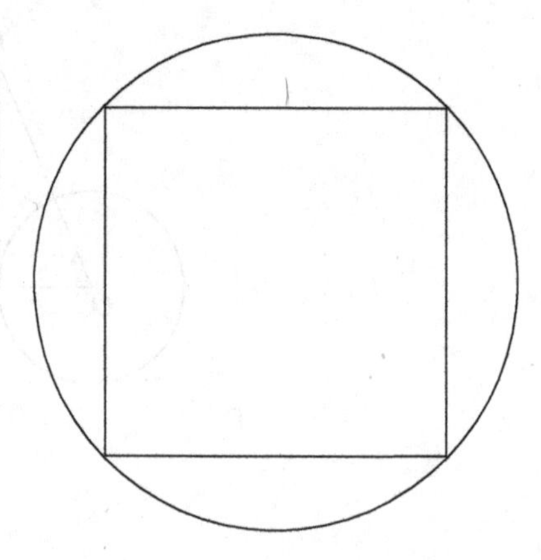

(A) $2\sqrt{2}$

(B) $3\sqrt{2}$

(C) $\pi\sqrt{2}$

(D) $2\pi\sqrt{2}$

(E) $3\pi\sqrt{2}$

26. A sphere and a cylinder both have the same volume (*V*) and the same radius (*r*). What is the height of the cylinder in terms of *r*?

(F) $4r$

(G) $\frac{3}{4}r$

(H) $\frac{4}{3}r$

(J) $\frac{\sqrt{3}}{3}r$

(K) $\frac{2\sqrt{3}}{3}r$

27. In the following graph, *a* is the amplitude of the wave, and *p* is its period. Which of the following is true?

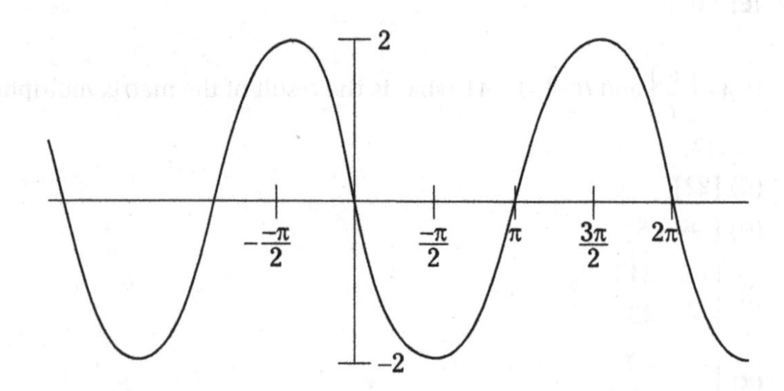

(A) $a = 2, p = \pi$

(B) $a = 2, p = 2\pi$

(C) $a = 4, p = \frac{1}{2}\pi$

(D) $a = 4, p = \pi$

(E) $a = 4, p = 2\pi$

28. If $\log_a b = 2$ and $a^2 = 100$, what is the value of b^2?

(F) $\sqrt{10}$

(G) 10

(H) 100

(J) 1,000

(K) 10,000

29. In the complex numbers, where $i^2 = -1$, what is the result when you multiply $\frac{1}{4+5i}$ by $\frac{2}{4-5i}$?

(A) $\frac{1}{8}$

(B) $-\frac{1}{8}$

(C) $\frac{2}{9}$

(D) $-\frac{2}{9}$

(E) $\frac{2}{41}$

30. In the following figure, the six points *A* through *F* are spaced equidistantly around a circle with a radius of 10. The intersection of ΔACE and ΔBDF is shaded. What is the area of this shaded region?

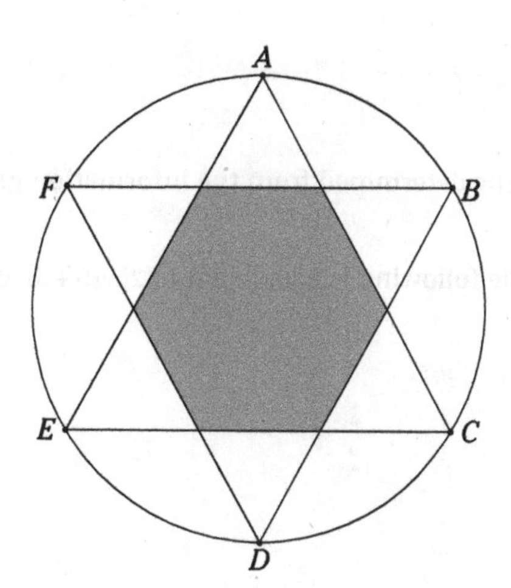

(F) $25\sqrt{3}$

(G) $50\sqrt{3}$

(H) $75\sqrt{3}$

(J) $125\sqrt{3}$

(K) $150\sqrt{3}$

31. Given that the law of cosines is $c^2 = a^2 + b^2 - 2ab\cos C$, what is the value of cos Q in the following triangle?

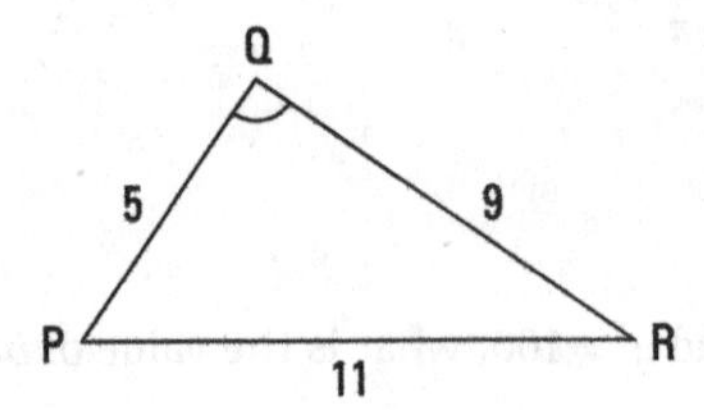

(A) $-\frac{1}{2}$

(B) $-\frac{1}{3}$

(C) $-\frac{1}{4}$

(D) $-\frac{1}{5}$

(E) $-\frac{1}{6}$

32. If $\log_k 256 = 2$, what is the value of $\sqrt{k}$?

(F) 2

(G) $2\sqrt{2}$

(H) 4

(J) $4\sqrt{2}$

(K) 8

33. What is the value of i^{987}?

(A) 1

(B) -1

(C) i

(D) $-i$

(E) Cannot be determined from the information given.

34. Which of the following is equivalent to $2\log 5 + 3\log 2 - \frac{1}{2}\log 4$?

(F) 1

(G) 2

(H) 3

(J) 4

(K) 5

35. In the following triangle, what is the value of $x°$, to the nearest whole degree?

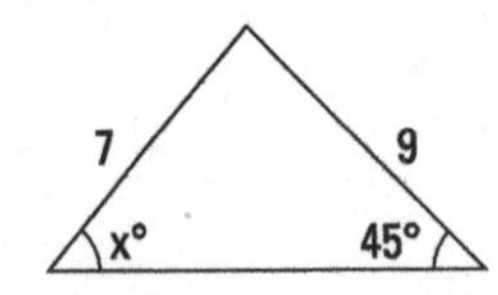

(A) 61°

(B) 62°

(C) 63°

(D) 64°

(E) 65°

36. Which of the following is equivalent to $\frac{(-2+4i)-(8-i)}{(5-3i)+(9-4i)}$?

(F) $\frac{5}{7}$

(G) $-\frac{5}{7}$

(H) $5i-7$

(J) $-\frac{5i}{7}$

(K) $-\frac{5i+1}{7}$

Solutions to Geometry, Trig, and Advanced Math Practice Problems

Here are the answers to the practice questions from the preceding section, complete with worked-out solutions.

1. C. The angles $a°$ and $b°$ are supplementary, so you can make the following equation:

$$a+b=180$$

The question tells you that $a°=2b°$, so substitute $2b$ for a in the preceding equation, and solve for b:

$$\begin{aligned}2b+b&=180\\3b&=180\\b&=60\end{aligned}$$

Thus, the value of c, f, and g is also 60, and the value of a, d, e, and h is 120. So

$$\begin{aligned}a+e&=120+120\\a+e&=240\end{aligned}$$

2. H. The area of ΔTUV is 34 and the base is $8+9=17$, so you can use the area formula for a triangle to find the height:

$$\begin{aligned}A&=\frac{1}{2}bh\\34&=\frac{1}{2}(17)h\\68&=17h\\4&=h\end{aligned}$$

The height of ΔTUV is also the height of ΔTUW, and the base of ΔTUW is 8, so

$$A = \frac{1}{2}bh = \frac{1}{2}(8)(4) = 16$$

3. E. The circle has an area of 11π, so use the area formula for a circle to find the radius:

$$11\pi = \pi r^2$$
$$11 = r^2$$
$$\sqrt{11} = r$$

The diameter of the circle is twice the radius, so it's $2\sqrt{11}$. This value is also the length of the side of the square, so use the perimeter formula for a square to get your answer:

$$P = 4s = 4(2\sqrt{11}) = 8\sqrt{11}$$

4. J. Recall that the tangent of an angle is the ratio of the opposite side over the adjacent side. So

$$\tan x = \frac{O}{A}$$

The opposite side is 1 and the adjacent side is 2, so plug these numbers in the preceding formula to get your answer:

$$= \frac{1}{2}$$

5. D. To calculate the area of the trapezoid, you need to know the height. To find it, draw a line in the original figure, like this:

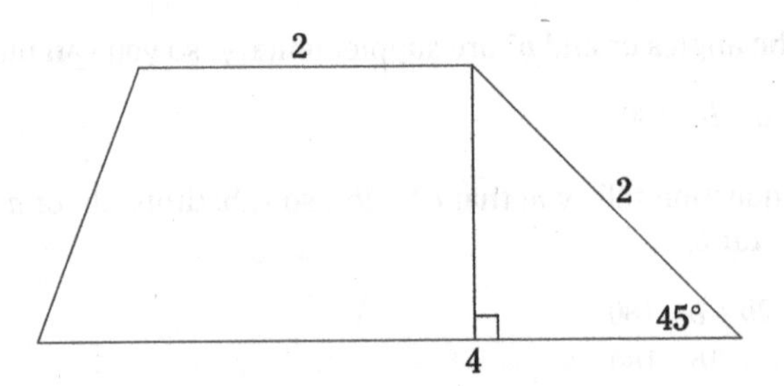

Note that the triangle is a 45-45-90 triangle with a hypotenuse of 2. The vertical leg of this triangle is also the height of the trapezoid. Recall that the hypotenuse of a 45-45-90 triangle is $\sqrt{2}$ times the leg. So calculate the length of a leg (x) as follows:

$$2 = x\sqrt{2}$$
$$\frac{2}{\sqrt{2}} = x$$

You can simplify this value, as I show you in Chapter 5, like this:

$$\frac{2}{\sqrt{2}} = \frac{2}{\sqrt{2}} \cdot \frac{\sqrt{2}}{\sqrt{2}} = \frac{2\sqrt{2}}{2} = \sqrt{2}$$

Thus, the length of each leg is $\sqrt{2}$. This is the height of the trapezoid, so plug this value into the formula for the area of a trapezoid, along with the lengths of the two bases, to find the answer:

$$A = \frac{b_1 + b_2}{2}h = \frac{2+4}{2}\sqrt{2} = 3\sqrt{2}$$

6. G. Begin by drawing a picture showing where Kailani's house, her school, and the mall are located:

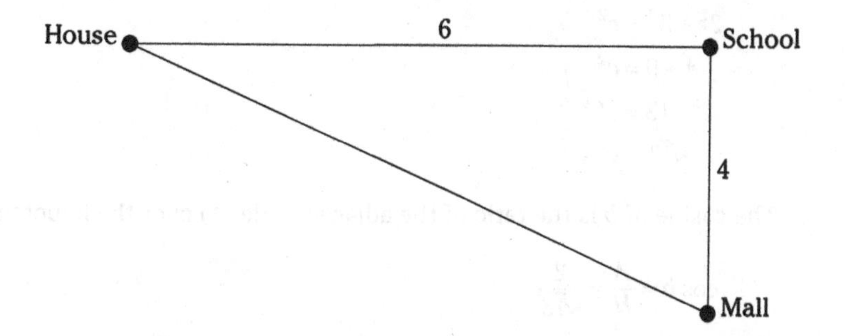

The three locations form a right triangle, so use the Pythagorean theorem $(a^2+b^2=c^2)$ to find the distance from the house to the mall:

$$\begin{aligned}4^2+6^2&=c^2\\16+36&=c^2\\52&=c^2\\c&=\sqrt{52}\approx 7\end{aligned}$$

7. E. First, use matrix addition to find the sum of the two matrices:

$$[3\quad 4\quad 1]+[5\quad -7\quad 0]=[8\quad -3\quad 1]$$

Now multiply the result by 4:

$$4[8\quad -3\quad 1]=[32\quad -12\quad 4]$$

8. J. The measures of the four interior angles in a quadrilateral add up to 360°, so you can make the following equation:

$$70+f+4f+(2f-25)=360$$

Solve for f:

$$\begin{aligned}70+f+4f+2f-25&=360\\45+7f&=360\\7f&=315\\f&=45\end{aligned}$$

9. D. The room has a square floor that has an area of 625 square feet, so calculate the length of one side of the room like this:

$$\begin{aligned}A&=s^2\\625&=s^2\\25&=s\end{aligned}$$

The volume of the room is 7,500 cubic feet, so find the height of the room using the formula for the volume of a box:

$$\begin{aligned}V&=lwh\\7{,}500&=(25)(25)h\\12&=h\end{aligned}$$

So the difference between the length of one side and the height of the room is $25-12=13$.

10. **J.** The figure shows a triangle with sides that are of lengths 2 and 3, so use the Pythagorean theorem to find the length of the hypotenuse:

$$\begin{aligned} 2^2 + 3^2 &= c^2 \\ 4 + 9 &= c^2 \\ 13 &= c^2 \\ \sqrt{13} &= c \end{aligned}$$

The cosine of b is the ratio of the adjacent side (2) over the hypotenuse ($\sqrt{13}$), so

$$\cos b = \frac{A}{H} = \frac{2}{\sqrt{13}}$$

To remove the radical from the denominator to get the correct answer, multiply both the numerator and the denominator by $\sqrt{13}$:

$$= \frac{2}{\sqrt{13}} \cdot \frac{\sqrt{13}}{\sqrt{13}} = \frac{2\sqrt{13}}{13}$$

11. **E.** The figure is a parallelogram, meaning the opposite sides are equal. So each of the two slanted sides has a length of 5. Thus, the triangle in the figure is a right triangle with one leg the length of 3 and the hypotenuse the length of 5. Using the Pythagorean theorem, you can find that $x = 4$. So the height of the parallelogram is 4, and the base is:

$$3 + 2x = 3 + 2(4) = 3 + 8 = 11$$

Use these values to calculate the area of the parallelogram:

$$A = bh = (11)(4) = 44$$

12. **K.** The circumference of the fountain is 24 feet, so use the formula for the circumference of a circle to find the radius:

$$\begin{aligned} C &= 2\pi r \\ 24 &= 2\pi r \\ \frac{12}{\pi} &= r \end{aligned}$$

The width of the walkway is 3 feet, so the radius of the outer edge of it is $\frac{12}{\pi} + 3$, which equals $\frac{12 + 3\pi}{\pi}$. Use this radius to find the circumference of the outer edge:

$$C = 2\pi r = 2\pi\left(\frac{12 + 3\pi}{\pi}\right) = 2(12 + 3\pi) = 24 + 6\pi$$

13. **E.** Change the logarithm to a power:

$$\log_3 v = 2 \quad \text{means} \quad 3^2 = v$$

Because $v = 9$, you know that $v^2 = 9^2 = 81$.

14. **G.** To solve this problem, use the arc length formula, plugging in 40° and a radius of 9:

$$\text{Arc length} = \text{degrees}\frac{\pi r}{180} = 40\frac{\pi(9)}{180} = \frac{360\pi}{180} = 2\pi$$

15. **E.** The two angles $(a+b)°$ and $(b-a-30)°$ are supplementary angles, so you can put together this equation:

$$a+b+b-a-30=180$$

Now simplify and solve for *b*:

$$\begin{aligned} 2b-30 &= 180 \\ 2b &= 210 \\ b &= 105 \end{aligned}$$

The two angles $(a+b)°$ and $(a+x)°$ are vertical angles, so

$$\begin{aligned} a+b &= a+x \\ b &= x \end{aligned}$$

Thus, $x=105$.

16. **K.** Use the formula for the surface area of a sphere to find the radius of a sphere with a surface area of 20:

$$\begin{aligned} A &= 4\pi r^2 \\ 20 &= 4\pi r^2 \\ \frac{5}{\pi} &= r^2 \\ \frac{\sqrt{5}}{\sqrt{\pi}} &= r \end{aligned}$$

Plug this value into the formula for the circumference of a circle:

$$C = 2\pi r = \frac{2\pi\sqrt{5}}{\sqrt{\pi}}$$

You now can simplify this answer by recognizing that $\pi = \sqrt{\pi}\sqrt{\pi}$:

$$= \frac{2\sqrt{\pi}\sqrt{\pi}\sqrt{5}}{\sqrt{\pi}} = 2\sqrt{\pi}\sqrt{5} = 2\sqrt{5\pi}$$

17. **A.** The problem tells you that $\sec\phi = y$, so use a trig identity to find $\cos\phi$:

$$\cos\phi = \frac{1}{\sec\phi} = \frac{1}{y}$$

The problem also tells you that $\sin\phi = x$, so use a trig identity to find $\tan\phi$:

$$\tan\phi = \frac{\sin\phi}{\cos\phi} = \frac{x}{\frac{1}{y}} = xy$$

18. F. You need to find the height in order to calculate the area, so draw in the height and split the base as follows:

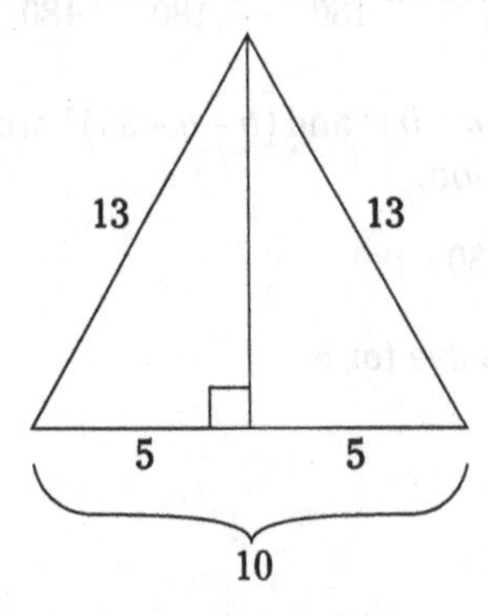

You can use the Pythagorean theorem to calculate the height, but you can save time by noticing that the triangle splits into two 5-12-13 triangles. So you know its height is 12. Thus, calculate its area using the formula for a triangle:

$$A = \frac{1}{2}bh = \frac{1}{2}(10)(12) = 60$$

19. E. Begin by FOILing the two complex numbers:

$$(1-i)(3+2i) = 3 + 2i - 3i - 2i^2$$

Next, combine like terms:

$$= 3 - i - 2i^2$$

As you know from the problem, $i^2 = -1$, so you can plug in that value:

$$= 3 - i - 2(-1)$$
$$= 3 - i + 2$$
$$= 5 - i$$

20. G. To begin, add an extra line to the figure, as follows:

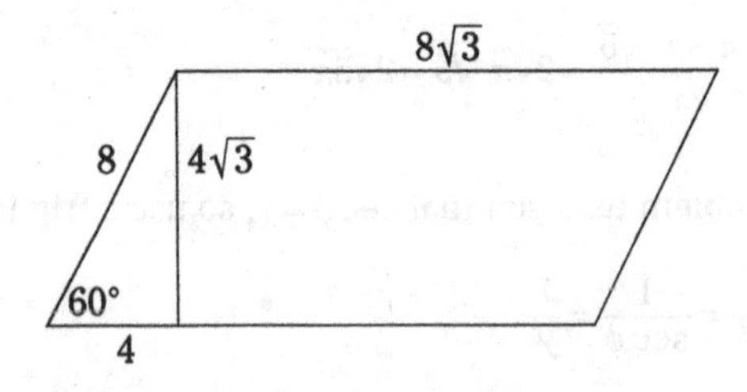

Notice that the resulting 30-60-90 triangle has a hypotenuse of 8, so it has legs with the lengths of 4 and $4\sqrt{3}$. So the height of the parallelogram is $4\sqrt{3}$ and its base is $8\sqrt{3}$. Plug these values into the formula for the area of a parallelogram:

$$A = bh = (8\sqrt{3})(4\sqrt{3}) = (32)(3) = 96$$

21. D. Use the formula for converting radians to degrees:

$$\frac{180}{\pi} = \frac{\text{degrees}}{\text{radians}}$$

Substitute $\frac{4\pi}{5}$ for radians and let d = degrees:

$$\frac{180}{\pi} = \frac{d}{\frac{4\pi}{5}}$$

To solve for d, begin by cross-multiplying. I suggest you do this in two steps:

$$180\frac{4\pi}{5} = \pi d$$

$$\frac{720\pi}{5} = \pi d$$

Simplify and solve:

$$144\pi = \pi d$$

$$144 = d$$

22. G. The area of the circle is 64π, so use the formula for the area of a circle to find the radius:

$$A = \pi r^2$$

$$64\pi = \pi r^2$$

$$64 = r^2$$

$$8 = r$$

Thus, $\overline{OQ} = 8$. $\overline{PQ}$ is tangent to the circle, so $\overline{PQ} \perp \overline{OQ}$. Thus, ΔPQO is a right triangle. Use the Pythagorean theorem:

$$c^2 = a^2 + b^2$$

$$c^2 = 8^2 + 15^2$$

$$c^2 = 64 + 225$$

$$c^2 = 289$$

$$c = 17$$

Thus, $\overline{OP} = 17$.

23. D. The long side of the triangle is $3\sqrt{3}$ and the hypotenuse is 6, so it's a 30-60-90 triangle as you see here:

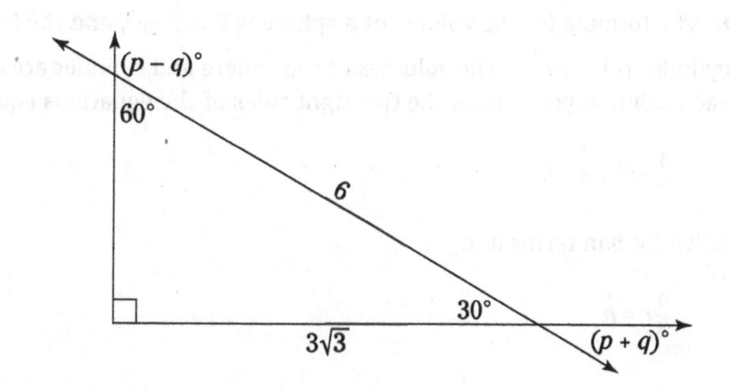

Because 30° and $(p+q)°$ are vertical angles, you can create this equation:

$$p+q=30$$

Additionally, 60° and $(p-q)°$ are supplementary angles, so

$$\begin{aligned} p-q+60&=180 \\ p-q&=120 \end{aligned}$$

Now you can add these two equations:

$$\begin{array}{r} p+q=30 \\ +\,p-q=120 \\ \hline 2p=150 \end{array}$$

Solve for *p*:

$$p=75$$

24. **J.** Make a chart using the two matrices in the order given in the multiplication:

$$\begin{array}{c|cc} & [-1 & 4] \\ \hline \begin{bmatrix}6\\7\end{bmatrix} & & \end{array}$$

Fill in the chart:

$$\begin{array}{c|cc} & [-1 & 4] \\ \hline \begin{bmatrix}6\\7\end{bmatrix} & \begin{matrix}-6\\-7\end{matrix} & \begin{matrix}24\\28\end{matrix} \end{array}$$

To get your answer, form a matrix according to the results in the chart:

$$\begin{bmatrix}-6 & 24\\-7 & 28\end{bmatrix}$$

25. **E.** The perimeter of the square is 12, so each side of the square is 3. Thus, the diagonal from one corner of the square to the other is $3\sqrt{2}$. This is also the diameter of the circle, so the radius of the circle is half of this, which is $\frac{3}{2}\sqrt{2}$. Plug this value into the formula for the circumference of a circle:

$$C=2\pi r=2\pi\left(\frac{3}{2}\sqrt{2}\right)=3\pi\sqrt{2}$$

26. **H.** The formula for the volume of a sphere is $V=\frac{4}{3}\pi r^3$, and the formula for the volume of a cylinder is $V=\pi r^2 h$. The volumes of the sphere and cylinder are equal, as are the radii of each solid, so you can set the two right sides of the equations equal:

$$\frac{4}{3}\pi r^3=\pi r^2 h$$

Solve for *h* in terms of *r*:

$$\frac{4}{3}r=h$$

27. B. The amplitude is the distance from the vertical middle of the graph to the crest, so $a = 2$. The period is the distance between two adjacent crests, so $p = 2\pi$. With this information, you can see that the correct answer is Choice (B).

28. K. Change the logarithm to a power:

$$\log_a b = 2 \quad \text{means} \quad a^2 = b$$

The question tells you that $a^2 = 100$, so $b = 100$. Therefore

$$b^2 = 100^2 = 10{,}000$$

29. E. Multiply the two fractions using the usual rules for multiplication:

$$\frac{1}{4+5i} \cdot \frac{2}{4-5i} = \frac{2}{(4+5i)(4-5i)}$$

FOIL the denominator and simplify:

$$= \frac{2}{16 - 20i + 20i - 25i^2} = \frac{2}{16 - 25i^2}$$

Substitute -1 for i^2 and solve:

$$= \frac{2}{16-(-25)} = \frac{2}{16+25} = \frac{2}{41}$$

30. G. To solve this problem, first find the length of the side of either of the large triangles. Note that both triangles are equilateral, so their angles measure 60° each. To find the length of one side, notice that the radius of the circle forms a 30-60-90 triangle as shown here:

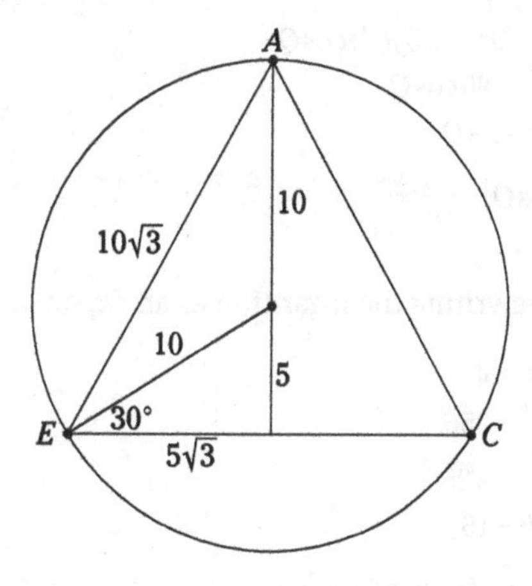

Thus, the length of the side of the triangle is $10\sqrt{3}$, and its height is 15. Plug these values into the formula for the area of a triangle:

$$A = \frac{1}{2}bh = \frac{1}{2}(10\sqrt{3})(15) = 75\sqrt{3}$$

Next, notice that each triangle separates into 9 small, equal-sized triangles:

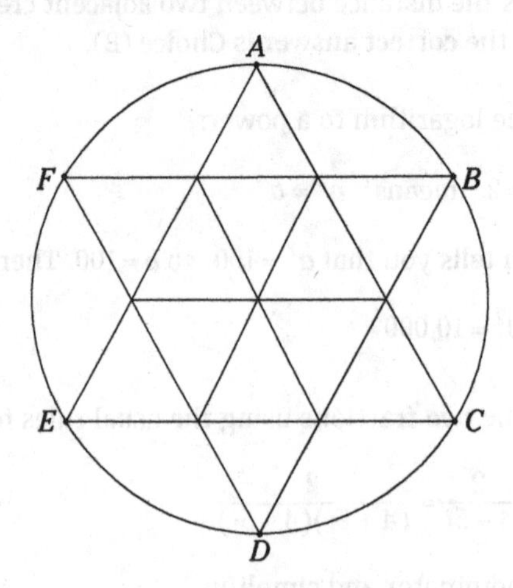

Each of these triangles has an area of $\frac{1}{9}$ the area of the larger triangle. So

$$\frac{75\sqrt{3}}{9} = \frac{25\sqrt{3}}{3}$$

The shaded area is composed of 6 of these small triangles, so its area is

$$6\left(\frac{25\sqrt{3}}{3}\right) = 50\sqrt{3}$$

31. E. Use the law of cosines, making sure to choose Q (the angle specified in the problem) as C:

$$c^2 = a^2 + b^2 - 2ab\cos C$$
$$11^2 = 5^2 + 9^2 - 2(5)(9)\cos Q$$
$$121 = 106 - 90\cos Q$$
$$15 = -90\cos Q$$
$$-\frac{1}{6} = \cos Q$$

32. H. Begin by rewriting the logarithm as an exponent:

$$\log_k 256 = 2$$
$$k^2 = 256$$
$$\sqrt{k^2} = \sqrt{256}$$
$$k = 16$$

Now, plug in 16 for k into $\sqrt{k}$:

$$\sqrt{k} = \sqrt{16} = 4$$

33. D. To begin, notice that $i^{987} \div 4 = 246$ r 3, so break i^{987} into two factors as follows:

$$i^{987} = i^{984} \cdot i^{3}$$

Now notice that $i^{984} = 1$, because 984 is divisible by 4:

$$= 1 \cdot i^{3}$$

From here, answer the question using your knowledge of exponents of i:

$$= i^{3} = -i$$

34. G. To begin, use $b \log_n a = \log_n a^b$ to rewrite the three terms in the expression. I do this in two steps to make sense of the process:

$$2\log 5 + 3\log 2 - \frac{1}{2}\log 4$$
$$= \log 5^2 + \log 2^3 - \log 4^{\frac{1}{2}}$$
$$= \log 25 + \log 8 - \log 2$$

Next, condense the first two terms down to one term using $\log_n a + \log_n b = \log_n ab$:

$$= \log 200 - \log 2.$$

Now, simplify this expression using $\log_n a - \log_n b = \log_n \frac{a}{b}$ and complete the problem:

$$= \log \frac{200}{2} = \log 100 = 2$$

35. E. Apply the law of sines, plugging in values for the two sides of the triangle and the angle that you know:

$$\frac{\sin A}{a} = \frac{\sin B}{b}$$
$$\frac{\sin 45°}{7} = \frac{\sin x}{9}$$
$$\frac{9 \sin 45°}{7} = \sin x$$

Now, use $\sin 45° = \frac{\sqrt{2}}{2}$, to simplify the left side of the equation, then solve for x using the inverse sine function on your calculator:

$$\sin x° = \frac{9}{7} \cdot \frac{\sqrt{2}}{2}$$
$$\sin x° \approx 0.909$$
$$\arcsin(\sin x°) = \arcsin 0.909$$
$$x° \approx 65°$$

36. G. To begin, simplify the numerator by adding and the denominator by subtracting:

$$\frac{(-2+4i)-(8-i)}{(5-3i)+(9-4i)} = \frac{-10+5i}{14-7i}$$

Factor out 5 from the numerator and −7 from the denominator, and then cancel out a factor of $-2+i$:

$$= \frac{5(-2+i)}{-7(-2+i)} = \frac{5}{-7} = -\frac{5}{7}$$

5 Tracking Your Progress with Practice Tests

IN THIS PART . . .

Timing yourself with two full-length, 60-minute math practice tests.

Checking your work with complete explanations for every question.

Chapter 14

Practice Test 1

Here's a practice test to help you prepare for the math portion of the ACT. The more practice you can get, the better. So take advantage of this opportunity and carve out another hour for this second practice test. Good luck!

In order to best simulate real exam conditions, I recommend doing the following:

1. **Sit where you won't be interrupted or tempted to pick up the TV remote or your phone.**
2. **Use the answer sheet provided to practice filling in the dots.**
3. **Set your timer for the time limits indicated at the beginning of the test.**
4. **Check your work for this test only; don't look at more than one test at a time.**
5. **Avoid taking a break during the test.**

TIP

When you finish this practice test, turn to Chapter 15, where you find detailed explanations of the answers as well as an abbreviated answer key. I recommend that you go through the answer explanations to all the questions, not just the ones that you missed, because you'll find lots of good info that may help you later.

Mathematics Test Answer Sheet

1. Ⓐ Ⓑ Ⓒ Ⓓ Ⓔ
2. Ⓕ Ⓖ Ⓗ Ⓙ Ⓚ
3. Ⓐ Ⓑ Ⓒ Ⓓ Ⓔ
4. Ⓕ Ⓖ Ⓗ Ⓙ Ⓚ
5. Ⓐ Ⓑ Ⓒ Ⓓ Ⓔ
6. Ⓕ Ⓖ Ⓗ Ⓙ Ⓚ
7. Ⓐ Ⓑ Ⓒ Ⓓ Ⓔ
8. Ⓕ Ⓖ Ⓗ Ⓙ Ⓚ
9. Ⓐ Ⓑ Ⓒ Ⓓ Ⓔ
10. Ⓕ Ⓖ Ⓗ Ⓙ Ⓚ
11. Ⓐ Ⓑ Ⓒ Ⓓ Ⓔ
12. Ⓕ Ⓖ Ⓗ Ⓙ Ⓚ
13. Ⓐ Ⓑ Ⓒ Ⓓ Ⓔ
14. Ⓕ Ⓖ Ⓗ Ⓙ Ⓚ
15. Ⓐ Ⓑ Ⓒ Ⓓ Ⓔ
16. Ⓕ Ⓖ Ⓗ Ⓙ Ⓚ
17. Ⓐ Ⓑ Ⓒ Ⓓ Ⓔ
18. Ⓕ Ⓖ Ⓗ Ⓙ Ⓚ
19. Ⓐ Ⓑ Ⓒ Ⓓ Ⓔ
20. Ⓕ Ⓖ Ⓗ Ⓙ Ⓚ
21. Ⓐ Ⓑ Ⓒ Ⓓ Ⓔ
22. Ⓕ Ⓖ Ⓗ Ⓙ Ⓚ
23. Ⓐ Ⓑ Ⓒ Ⓓ Ⓔ
24. Ⓕ Ⓖ Ⓗ Ⓙ Ⓚ
25. Ⓐ Ⓑ Ⓒ Ⓓ Ⓔ
26. Ⓕ Ⓖ Ⓗ Ⓙ Ⓚ
27. Ⓐ Ⓑ Ⓒ Ⓓ Ⓔ
28. Ⓕ Ⓖ Ⓗ Ⓙ Ⓚ
29. Ⓐ Ⓑ Ⓒ Ⓓ Ⓔ
30. Ⓕ Ⓖ Ⓗ Ⓙ Ⓚ
31. Ⓐ Ⓑ Ⓒ Ⓓ Ⓔ
32. Ⓕ Ⓖ Ⓗ Ⓙ Ⓚ
33. Ⓐ Ⓑ Ⓒ Ⓓ Ⓔ
34. Ⓕ Ⓖ Ⓗ Ⓙ Ⓚ
35. Ⓐ Ⓑ Ⓒ Ⓓ Ⓔ
36. Ⓕ Ⓖ Ⓗ Ⓙ Ⓚ
37. Ⓐ Ⓑ Ⓒ Ⓓ Ⓔ
38. Ⓕ Ⓖ Ⓗ Ⓙ Ⓚ
39. Ⓐ Ⓑ Ⓒ Ⓓ Ⓔ
40. Ⓕ Ⓖ Ⓗ Ⓙ Ⓚ
41. Ⓐ Ⓑ Ⓒ Ⓓ Ⓔ
42. Ⓕ Ⓖ Ⓗ Ⓙ Ⓚ
43. Ⓐ Ⓑ Ⓒ Ⓓ Ⓔ
44. Ⓕ Ⓖ Ⓗ Ⓙ Ⓚ
45. Ⓐ Ⓑ Ⓒ Ⓓ Ⓔ
46. Ⓕ Ⓖ Ⓗ Ⓙ Ⓚ
47. Ⓐ Ⓑ Ⓒ Ⓓ Ⓔ
48. Ⓕ Ⓖ Ⓗ Ⓙ Ⓚ
49. Ⓐ Ⓑ Ⓒ Ⓓ Ⓔ
50. Ⓕ Ⓖ Ⓗ Ⓙ Ⓚ
51. Ⓐ Ⓑ Ⓒ Ⓓ Ⓔ
52. Ⓕ Ⓖ Ⓗ Ⓙ Ⓚ
53. Ⓐ Ⓑ Ⓒ Ⓓ Ⓔ
54. Ⓕ Ⓖ Ⓗ Ⓙ Ⓚ
55. Ⓐ Ⓑ Ⓒ Ⓓ Ⓔ
56. Ⓕ Ⓖ Ⓗ Ⓙ Ⓚ
57. Ⓐ Ⓑ Ⓒ Ⓓ Ⓔ
58. Ⓕ Ⓖ Ⓗ Ⓙ Ⓚ
59. Ⓐ Ⓑ Ⓒ Ⓓ Ⓔ
60. Ⓕ Ⓖ Ⓗ Ⓙ Ⓚ

Mathematics Test

TIME: 60 minutes for 60 questions

DIRECTIONS: Each question has five answer choices. Choose the best answer for each question, and then shade in the corresponding oval on your answer sheet.

1. Jackson worked 25 hours and received \$225. At the same rate of pay, how much would he make if he worked 40 hours?

(A) \$300

(B) \$325

(C) \$350

(D) \$360

(E) \$400

2. What is the missing number in the sequence 1, 5, 10, 16, 23, 31, ___?

(F) 37

(G) 38

(H) 39

(J) 40

(K) 41

3. If $x = 6$ and $y = -2$, what is the value of $3xy + 2x^2 - y^3$?

(A) 44

(B) 48

(C) 50

(D) 52

(E) 56

4. Noreen recently took a job helping people register to vote. The job has a mandatory 10-day period of probation during which her success rate is strictly monitored. On her first day, she registered 30 people. Then, for each of the next 9 days, she registered 4 more people than she did on the previous day. How many people did she register altogether during her probationary period?

(F) 300

(G) 340

(H) 480

(J) 560

(K) 600

5. In the following figure, the circle centered at N has a radius of 4. What is the area of the shaded region?

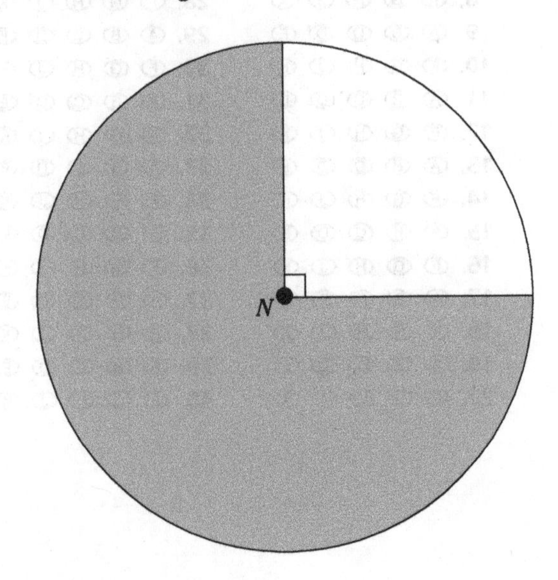

(A) 3π

(B) 6π

(C) 9π

(D) 12π

(E) 16π

6. Latisha's three bowling scores in a tournament were 167, 178, and 186. What was her average score for the tournament?

(F) 176

(G) 177

(H) 178

(J) 179

(K) 180

7. What is the value of k if $\sqrt{10k+3} = 5$?

(A) 0.2

(B) 2

(C) 2.2

(D) 2.8

(E) 4.7

8. Which of the following is equivalent to $-8(x-2)<3x-6$?

(F) $x<2$

(G) $x>2$

(H) $x\geq -2$

(J) $x<-2$

(K) $x>-2$

9. How many different positive integers are factors of both 28 and 42?

(A) 1

(B) 2

(C) 3

(D) 4

(E) More than 4

10. What is the slope of a line that includes the points $(-4,1)$ and $(10,-6)$?

(F) 2

(G) $\frac{1}{2}$

(H) $-\frac{1}{2}$

(J) $\frac{6}{7}$

(K) $-\frac{7}{6}$

11. If $3x+5y=4$, which of the following is equivalent to the expression $(6x+10y)(100x+100y)$?

(A) $100x+100y$

(B) $200x+200y$

(C) $400x+400y$

(D) $800x+800y$

(E) $1{,}600x+1{,}600y$

12. What is the value of x if $x^2-5x-14=0$ and $x>0$?

(F) 2

(G) 4

(H) 5

(J) 7

(K) 9

13. If $\frac{3n}{2}=\frac{4n+3}{3}$, then $n=$

(A) 6

(B) 7

(C) 9

(D) 11

(E) 13

14. If the height of an equilateral triangle is 9, what is its area?

(F) 27

(G) 54

(H) 81

(J) $27\sqrt{3}$

(K) $54\sqrt{3}$

15. In the following figure, what is the value of y in terms of x?

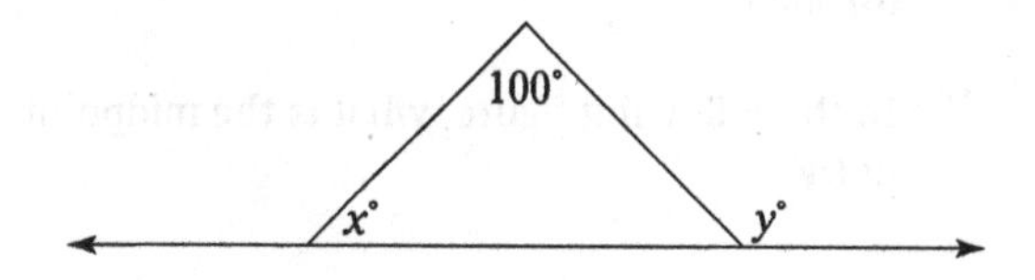

(A) $x+80$

(B) $80-x$

(C) $x+100$

(D) $x-100$

(E) $100-x$

16. If 15% of n is 300, what is 22% of n?

(F) 400

(G) 440

(H) 480

(J) 525

(K) 550

GO ON TO NEXT PAGE

17. What is the formula of a line that is perpendicular to $y = \frac{1}{3}x + 9$ and includes the point (3, 4)?

(A) $y = \frac{1}{3}x + 5$

(B) $y = -\frac{1}{3}x + 13$

(C) $y = 3x + 5$

(D) $y = -3x + 5$

(E) $y = -3x + 13$

18. Two values of m satisfy the equation $|5m - 11| - 3m = 9$. What is the result when you multiply these two values together?

(F) 2.5

(G) 2.75

(H) 3.2

(J) 3.75

(K) 4.25

19. In the following figure, what is the midpoint of $\overline{UV}$?

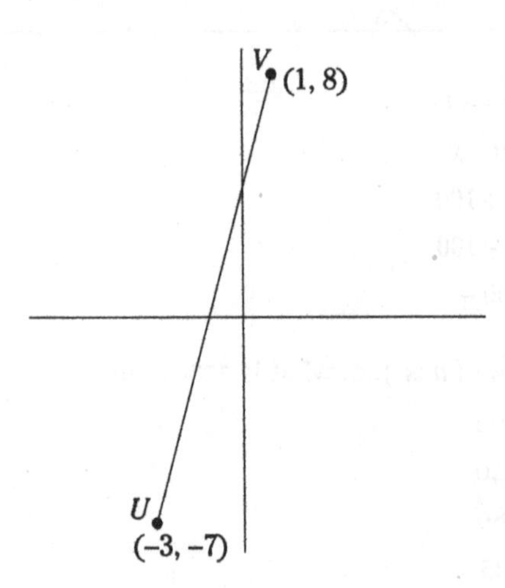

(A) $\left(-1, \frac{1}{2}\right)$

(B) $\left(-1, \frac{3}{2}\right)$

(C) $\left(-1, -\frac{1}{2}\right)$

(D) $\left(-2, \frac{1}{2}\right)$

(E) $\left(-2, -\frac{3}{2}\right)$

20. If $f(x) = x^2 + 9$ and $g(x) = 24 + 4x$, what is the value of $\frac{f(4)}{g(-1)}$?

(F) 0.75

(G) 0.8

(H) 1.2

(J) 1.25

(K) 1.75

21. In the following figure, what is the value of x in terms of y?

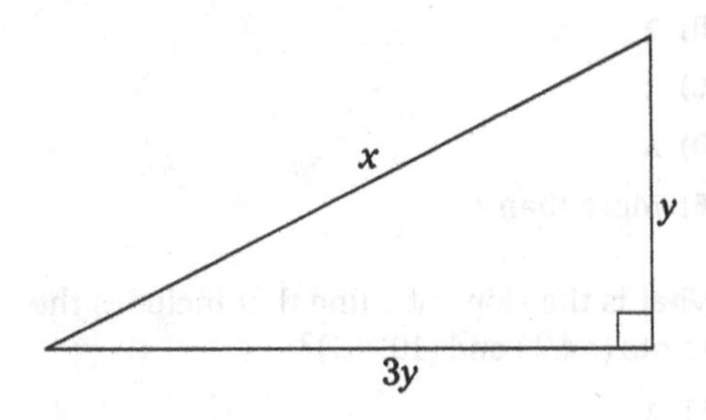

(A) $10y$

(B) $10y^2$

(C) $y\sqrt{10}$

(D) $\sqrt{10y}$

(E) $10\sqrt{y}$

22. Two variables, v and w, are inversely proportional such that when $v = 7$, $w = 14$. What is the value of w when $v = 2$?

(F) 1

(G) 4

(H) 14

(J) 28

(K) 49

23. The ratio of adults to girls to boys on a class field trip was 1:4:5. If the trip included 6 more boys than girls, how many adults were with the group?

(A) 3

(B) 4

(C) 6

(D) 8

(E) 12

24. In the following figure, line a and line b are parallel and pass through the points shown. What is the equation for line b?

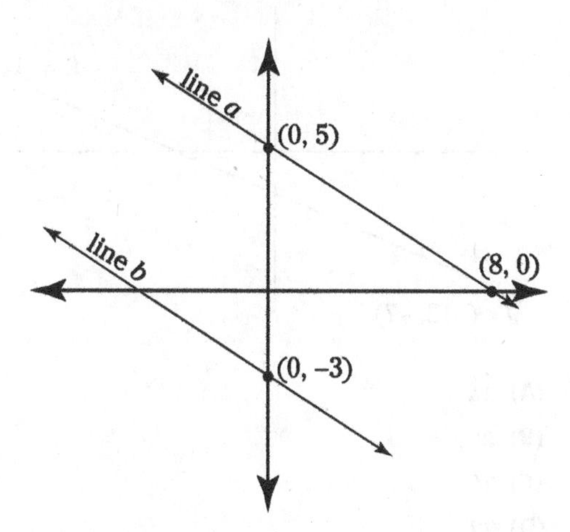

(F) $y = \frac{5}{8}x - 3$

(G) $y = -\frac{5}{8}x + 5$

(H) $y = -\frac{5}{8}x - 3$

(J) $y = \frac{8}{5}x + 5$

(K) $y = -\frac{8}{5}x - 3$

25. A bag contains 7 black socks, 12 white socks, and 17 red socks. If you pick one sock at random from the bag, what is the probability that it will NOT be white?

(A) $\frac{1}{2}$

(B) $\frac{2}{3}$

(C) $\frac{3}{4}$

(D) $\frac{7}{12}$

(E) $\frac{7}{36}$

26. If $\tan\theta = \frac{4}{3}$, then $\sin\theta =$

(F) $\frac{3}{4}$

(G) $\frac{3}{5}$

(H) $\frac{4}{5}$

(J) $\frac{5}{3}$

(K) $\frac{5}{4}$

27. If $pq = 3$, then $p^3q^4 + p^4q^5 =$

(A) $12q$

(B) $7p + 9q$

(C) $12p + 20q$

(D) $96p$

(E) $108q$

28. Jane ran around the perimeter of a rectangular park at a constant rate of 10 feet per second. The park has an area of 67,500 square feet, and its length is exactly three times its width. For how many seconds did Jane run?

(F) 60

(G) 120

(H) 240

(J) 360

(K) 480

Use this information to answer Questions 29 and 30: Danielle's phone plan charges her \$30 per month for the first 200 minutes and then \$0.10 per minute for each subsequent minute.

---

29. Which of the following functions takes an input of any whole-number value of $x \geq 200$ and outputs the value for $f(x)$ as the amount of dollars Danielle would pay for x minutes of phone usage?

(A) $f(x) = 0.1x + 10$

(B) $f(x) = 0.1x + 20$

(C) $f(x) = 0.1x + 30$

(D) $f(x) = 0.1x + 200$

(E) $f(x) = 0.1x + 230$

30. If Danielle paid exactly \$100 last month, how many minutes did she use?

(F) 700

(G) 800

(H) 900

(J) 1,000

(K) 1,200

GO ON TO NEXT PAGE

31. What is the area of ΔRST in the following figure?

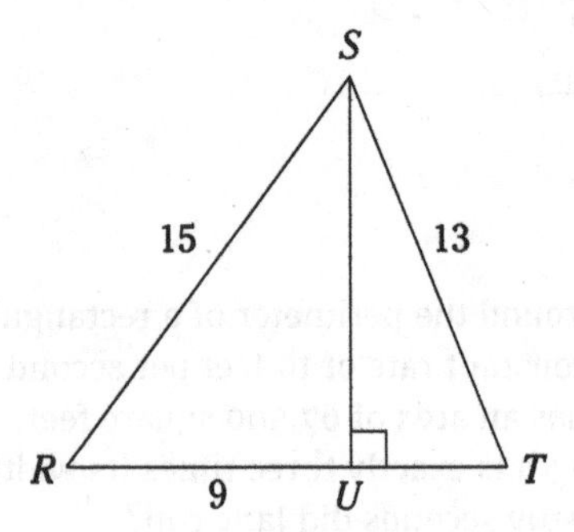

(A) 84

(B) 91

(C) 96

(D) 105

(E) 120

32. Antoine bought a new electric guitar that cost $588.60 after 9% sales tax was added. What was the price of the guitar without tax?

(F) $536

(G) $540

(H) $542

(J) $545

(K) $548

33. Which of the following points on the *xy*-graph is the *x*-intercept of the equation $y = 2x - 8$?

(A) (0, 4)

(B) $\left(0, \frac{1}{4}\right)$

(C) (4, 0)

(D) $(-4, 0)$

(E) $\left(-\frac{1}{4}, 0\right)$

34. What is the determinant of the matrix $\begin{bmatrix} 3 & 6 \\ -1 & 2 \end{bmatrix}$?

(F) 0

(G) 12

(H) [0]

(J) [6]

(K) [12]

35. In the following figure, what is the length of $\overline{JK}$?

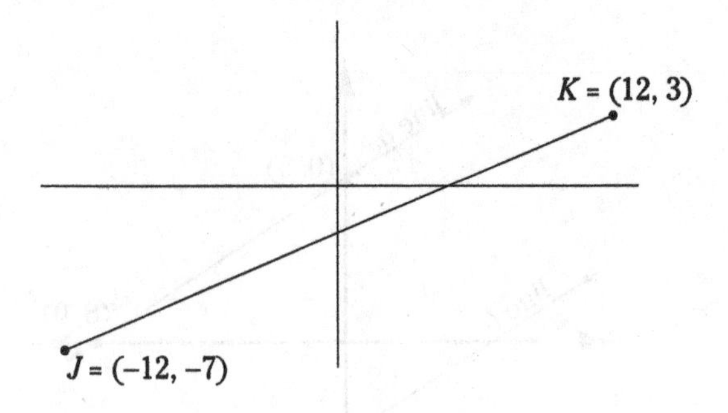

(A) 24

(B) 25

(C) 26

(D) 27

(E) 28

36. If $\left(\frac{1}{49}\right)^{n+3} = \sqrt{7}$, then $n =$

(F) $-\frac{7}{4}$

(G) $\frac{9}{4}$

(H) $-\frac{11}{4}$

(J) $-\frac{13}{4}$

(K) $\frac{15}{4}$

37. A 25-foot ladder stands against a vertical wall at an angle of *n* degrees with the ground. If $\sin n = \frac{4}{5}$, how far is the base of the ladder from the wall?

(A) 12

(B) 13

(C) 14

(D) 15

(E) 16

38. In the following figure, if the dimensions of the trapezoid are as shown and the area of the trapezoid is 144, what is the value of x?

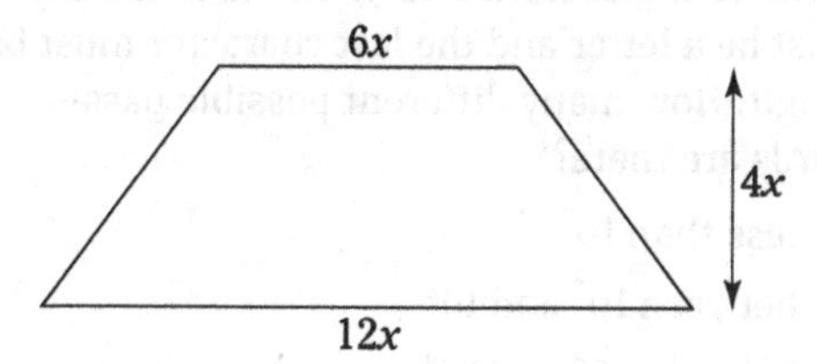

(F) 2

(G) 3

(H) 4

(J) 6

(K) 8

39. Ansgar is writing a novel. He writes seven days a week. On each of those days he writes for at least 4 hours but never more than 8 hours. Last week, he wrote for exactly 46 hours. What is the maximum number of days on which he could have written for 8 hours?

(A) 2 days

(B) 3 days

(C) 4 days

(D) 5 days

(E) 6 days

40. Which of the following is a possible value of x if $5x^2 - 10x + 4 = 0$?

(F) $2\sqrt{5}$

(G) $1+2\sqrt{5}$

(H) $1+\frac{\sqrt{5}}{5}$

(J) $2+\frac{\sqrt{5}}{10}$

(K) $10+\frac{\sqrt{5}}{10}$

41. If you multiply a number by 3 and then add 40, the result is the same as if you first add 17 and then multiply by 2. What is the result if you subtract 9 from the number and then multiply by 4?

(A) –60

(B) –72

(C) –84

(D) –108

(E) –124

42. If $7x + 4y = 18$ and $3x + y = -3$, what is the value of $x + y$?

(F) 9

(G) 11

(H) 12

(J) 14

(K) 15

43. In the following figure, the area of the shaded region is 20% of the area of the whole circle centered at P. The angle shown measures d degrees. What is its measurement in radians?

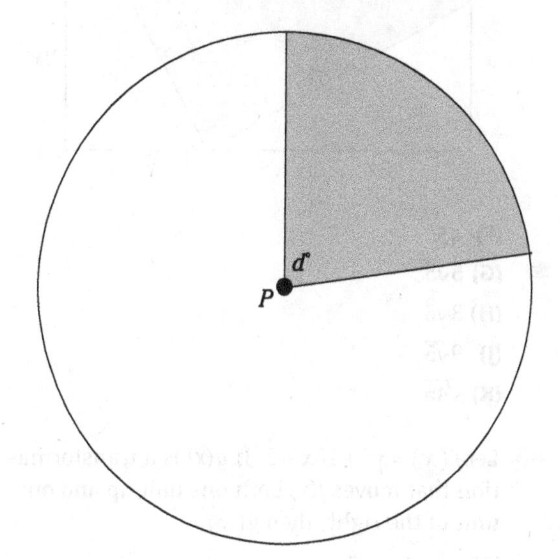

(A) $\frac{1}{5}\pi$

(B) $\frac{2}{5}\pi$

(C) $\frac{4}{5}\pi$

(D) $\frac{2}{15}\pi$

(E) $\frac{4}{15}\pi$

GO ON TO NEXT PAGE

44. In the following figure, the area of the large square is 81. What is the area of the shaded square?

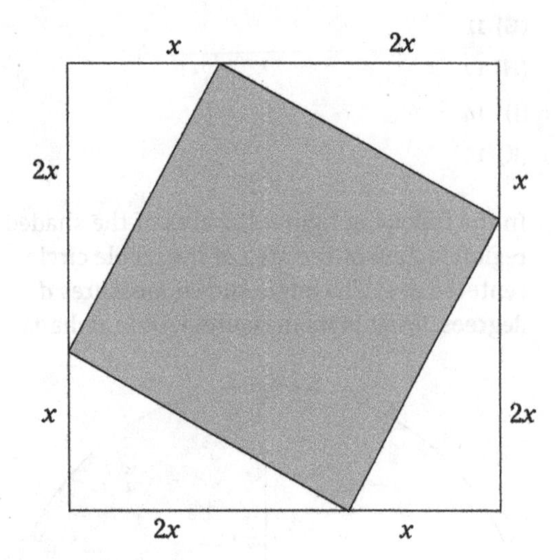

(F) 45

(G) $5\sqrt{3}$

(H) $3\sqrt{5}$

(J) $9\sqrt{5}$

(K) $\sqrt{45}$

45. Let $f(x) = x^2 + 10x + 2$. If $g(x)$ is a transformation that moves $f(x)$ both one unit up and one unit to the right, then $g(x) =$

(A) $x^2 + 8x - 6$

(B) $x^2 + 9x + 3$

(C) $x^2 + 10x - 6$

(D) $x^2 + 11x + 3$

(E) $x^2 + 12x + 6$

46. If $\frac{a+b}{10} = \frac{a - 0.1b^2}{a-b}$, what is the value of a?

(F) 0.01

(G) 0.1

(H) 1

(J) 10

(K) 100

47. A password for a computer system requires exactly 6 characters. Each character can be either one of the 26 letters from A to Z or one of the 10 digits from 0 to 9. The first character must be a letter and the last character must be a digit. How many different possible passwords are there?

(A) less than 10^7

(B) between 10^7 and 10^8

(C) between 10^8 and 10^9

(D) between 10^9 and 10^{10}

(E) more than 10^{10}

48. On the *xy*-plane, what is the area of a circle with this equation: $(x+3)^2 + (y-4)^2 = 49$?

(F) 5π

(G) 7π

(H) 25π

(J) 49π

(K) 125π

49. Which of the following is equal to $\sin x \sec x$?

(A) $\tan x$

(B) $\cot x$

(C) $\cos x \tan x$

(D) $\cos x \csc x$

(E) $\cot x \csc x$

50. The following figure shows a cylindrical tank whose diameter is 3 times the length of its height. The tank holds approximately 231.5 cubic meters of fluid. Which of the following answer choices most closely approximates the height of the tank?

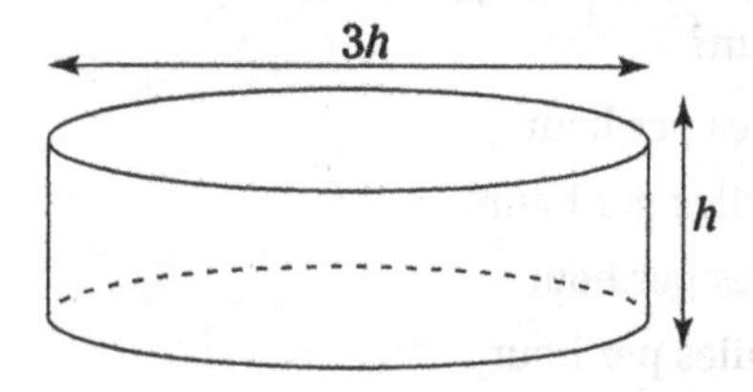

(F) 2 meters

(G) 3 meters

(H) 4 meters

(J) 5 meters

(K) 6 meters

51. Paulette, Quentin, and Rosie each donated money to a charity. Paulette gave as much money as Quentin and Rosie gave together. If Quentin had given three times more than he gave, he would have given \$40 more than Paulette. And if Rosie had given \$20 less, she would have given half as much as Paulette. How much did Paulette give?

(A) \$80

(B) \$120

(C) \$160

(D) \$200

(E) \$240

To answer Questions 52 and 53, use the following graph, which provides information about the number of new clients five salespeople registered last month.

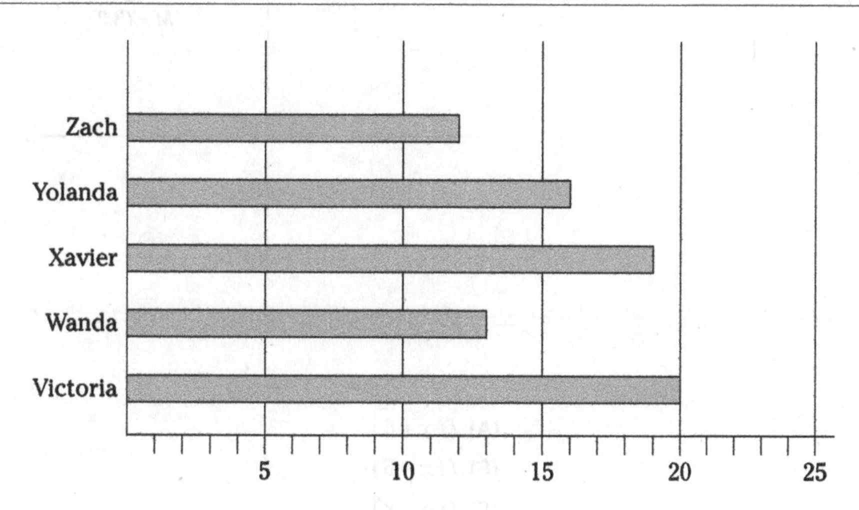

52. What percentage of the new clients did Yolanda register?

(F) 18%

(G) 20%

(H) 22%

(J) 24%

(K) 25%

GO ON TO NEXT PAGE

53. Suppose that next month Victoria registers twice as many clients as she did this month and that each of the other four salespeople registers the same number of clients as they did this month. In this case, what percentage of clients will Victoria have registered?

(A) 36%

(B) 40%

(C) 44%

(D) 48%

(E) 50%

54. In the following figure, the regular octagon has a side with a length of 1. What is the area of the shaded region?

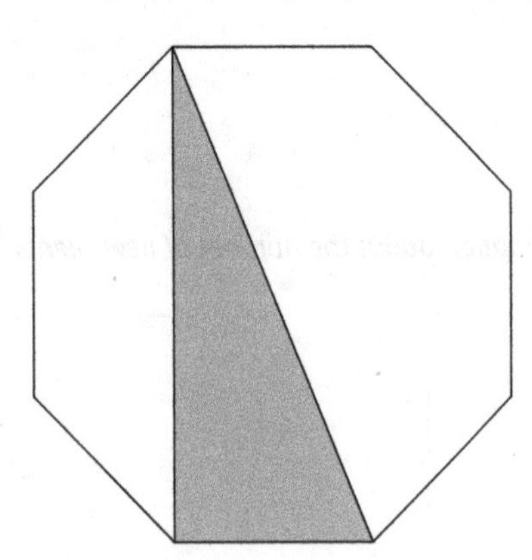

(F) $\sqrt{2}+1$

(G) $\sqrt{2}+2$

(H) $\frac{\sqrt{2}+1}{2}$

(J) $\frac{\sqrt{2}}{2}+1$

(K) $\frac{\sqrt{2}}{2}+2$

55. If $\frac{a}{c}-\frac{a}{b}=\frac{b-c}{a}$, with $a>0$, $b>0$, and $c>0$, what is the value of a in terms of b and c?

(A) $b-c$

(B) $\sqrt{bc}$

(C) $\sqrt{b-c}$

(D) $\frac{\sqrt{b-c}}{bc}$

(E) $\frac{\sqrt{bc}}{b-c}$

56. At 10:00, Angela starts from her home and runs at a constant pace to Kathleen's house, which is exactly 2 miles away. Immediately, she and Kathleen turn around and walk back to Angela's house exactly 4 miles an hour slower than Angela ran. When they arrive at Angela's house, the time is 10:45. At what speed did Angela run?

(F) 6 miles per hour

(G) 6.5 miles per hour

(H) 7 miles per hour

(J) 7.5 miles per hour

(K) 8 miles per hour

57. The following figure shows $f(x)$, which includes points L, M, and N plus the line segments $\overline{LM}$ and $\overline{MN}$. Which of the following functions is equivalent to $f(x)$?

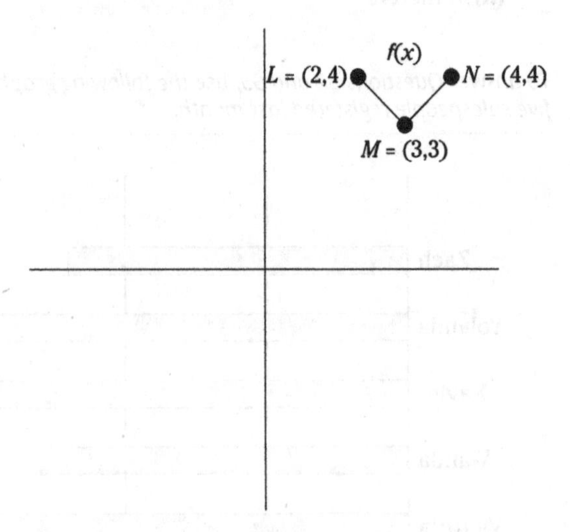

(A) $f(x+6)$

(B) $f(x-6)$

(C) $f(6-x)$

(D) $-f(x+6)$

(E) $-f(x-6)$

58. If $\log_9 n = \frac{1}{2}$ and $n > 0$, what is the value of $\sqrt{n}$?

(F) 3

(G) $\sqrt{3}$

(H) $\sqrt[3]{3}$

(J) $\sqrt{8}$

(K) $\sqrt[3]{9}$

59. In the complex numbers, where $i = \sqrt{-1}$, the conjugate of any value $a + bi$ is $a - bi$. What is the result when you multiply $2 + 7i$ by its conjugate?

(A) 45

(B) –45

(C) $45i$

(D) 53

(E) $53i$

60. Mo works as a lifeguard at a local pool. At the beginning of a 12-hour overnight shift, the pool was full, and Mo began draining it. After 2 hours, the pool was completely empty. He spent 3 hours cleaning the pool and then began filling it up again. The pool finished filling just as his shift ended. Which of the following graphs accurately describes the amount of water in the pool throughout Mo's shift?

(F)

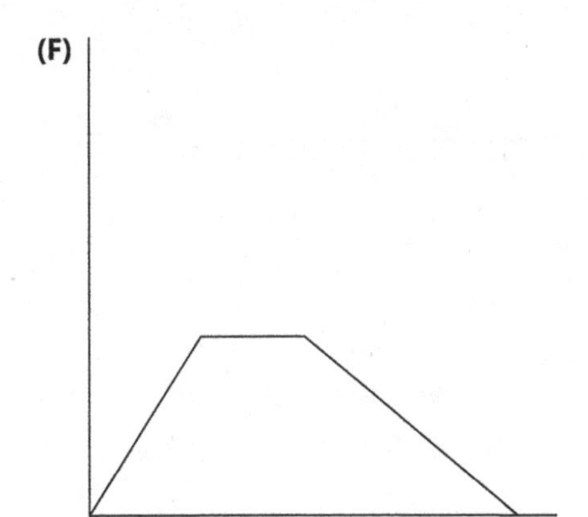

(G)

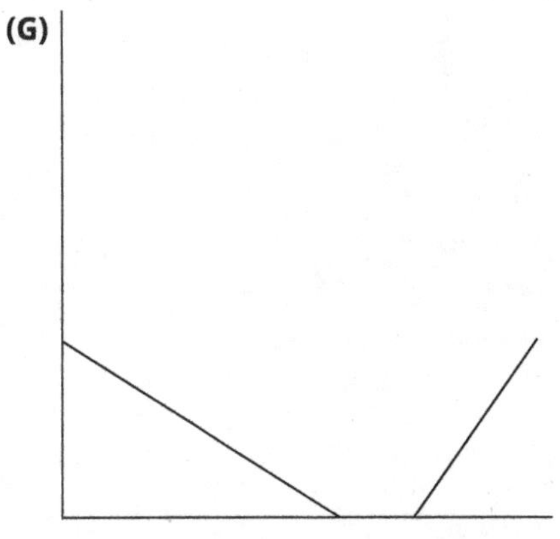

(H)

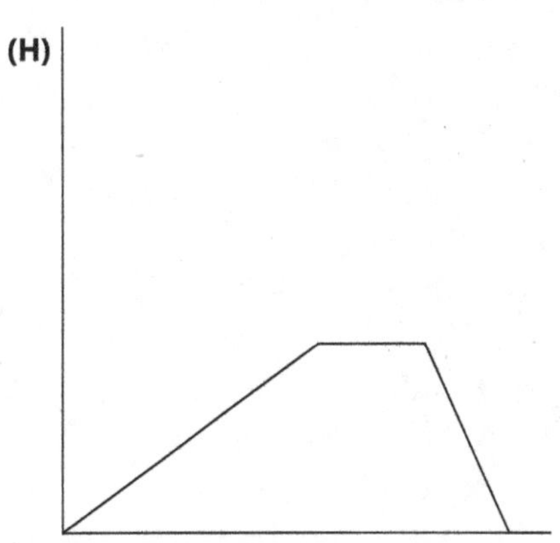

(J)

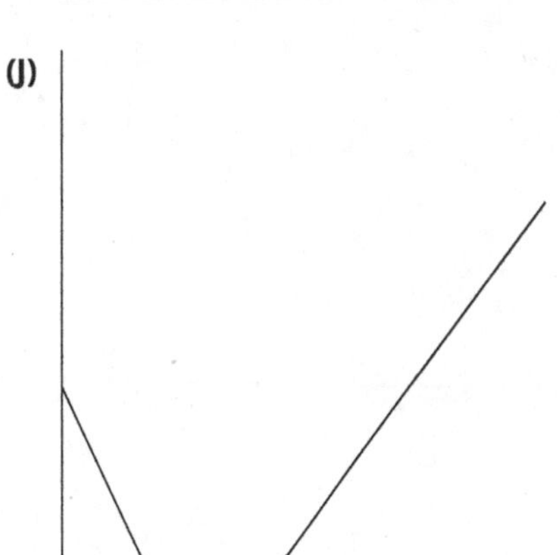

(K)

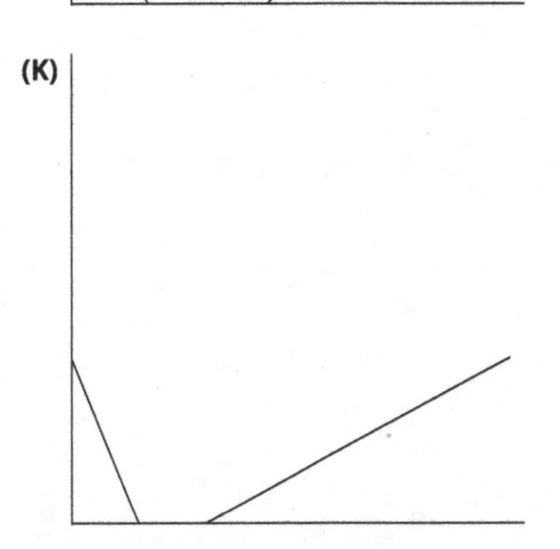

DO NOT TURN THE PAGE UNTIL TOLD TO DO SO **STOP** DO NOT RETURN TO A PREVIOUS TEST

Chapter 15
Practice Test 1: Answers and Explanations

So you've completed Practice Test 1? I bet you feel more prepared already. Use this chapter to check your answers and find out just how prepared you are. In this chapter, I provide detailed explanations along with a bare-bones answer key. The answer key is great for when you're checking your answers in a hurry, but I suggest that you review the detailed explanations for each question — even those you answered correctly. The explanations show you how to calculate each problem and often provide tips and tricks.

Mathematics Test

1. **D.** Jackson worked 25 hours for $225, so divide to find out how much he earned per hour:

   $225 \div 25 = 9$

   Now multiply $9 per hour by 40 hours:

   $40 \times 9 = 360$

   Therefore, he would earn $360 in 40 hours.

2. **J.** Each number in the sequence is a bit higher than the one before it, so see how much needs to be added to each number to produce the rest. As you can see, if you add 4, then 5, then 6, and so forth, the numbers add up correctly:

   $1(+4=)5(+5=)10(+6=)16(+7=)23(+8=)31(+9=)40$

3. **A.** First, plug in 6 for *x* and –2 for *y* throughout the expression:

$$3xy + 2x^2 - y^3 = 3(6)(-2) + 2(6)^2 - (-2)^3$$

Now simplify using the order of operations:

$$= 3(6)(\ -2) + 2(6)(6) - (-2)(-2)(-2) = -36 + 72 + 8 = 44$$

4. **H.** Jot down how many people Noreen registered on each day:

30	34	38	42	46	50	54	58	62	66

To save time adding all these numbers, notice that the total of the first and 10th numbers is $30 + 66 = 96$. This total is the same for the 2nd and 9th, the 3rd and 8th, the 4th and 7th, and the 5th and 6th. Therefore, you have 5 pairings of days on which Noreen registered 96 people. You can simply multiply to find the total:

$$96 \times 5 = 480$$

5. **D.** Use the formula for the area of a circle:

$$A = \pi r^2 = \pi(4)^2 = 16\pi$$

The right angle accounts for 90° of the 360° circle, which is $\frac{1}{4}$ of it. So the shaded region of the circle is $\frac{3}{4}$ the area of the circle:

$$\frac{3}{4}(16\pi) = 12\pi$$

6. **G.** Latisha's scores for 3 games were 167, 178, and 186. To find the average, simply place these numbers into the formula for the mean:

$$\text{Mean} = \frac{\text{Sum of values}}{\text{Number of values}} = \frac{167 + 178 + 186}{3} = \frac{531}{3} = 177$$

7. **C.** Start by squaring both sides of the equation to undo the square root:

$$\sqrt{10k + 3} = 5$$
$$10k + 3 = 25$$

Now solve for *k*:

$$10k = 2$$
$$2k = 2.2$$

8. **G.** Distribute the left side and combine like terms:

$$-8(x - 2) < 3x - 6$$
$$-8x + 16 < 3x - 6$$
$$-11x + 16 < -6$$
$$-11x < -22$$

To solve for *x*, divide both sides by -11 and reverse the inequality:

$$x > 2$$

9. **D.** List the factors of both 28 and 42:

Factors of 28: 1	2	4	7	14	28		
Factors of 42: 1	2	3	6	7	14	21	42

Now you can see that 1, 2, 7, and 14 are factors of both numbers, so the correct answer is Choice (D).

10. **H.** Plug the values $(-4, 1)$ and $(10, -6)$ into the two-point slope formula:

$$\text{Slope} = \frac{y_2 - y_1}{x_2 - x_1} = \frac{-6-1}{10-(-4)} = \frac{-7}{14} = -\frac{1}{2}$$

11. **D.** To begin, factor a 2 out of $(6x + 10y)$:

$$(6x + 10y)(100x + 100y) = 2(3x + 5y)(100x + 100y)$$

Now substitute 4 for $3x + 5y$ and distribute:

$$2(4)(100x + 100y) = 8(100x + 100y) = 800x + 800y$$

12. **J.** Factor the left side of the equation:

$$x^2 - 5x - 14 = 0$$
$$(x + 2)(x - 7) = 0$$

Next, split the equation into two separate equations and solve each for x:

$$x + 2 = 0 \qquad x - 7 = 0$$
$$x = -2 \qquad x = 7$$

Because $x > 0$, the value of x is 7.

13. **A.** Cross-multiply to get rid of the fractions:

$$\frac{3n}{2} = \frac{4n+3}{3}$$
$$3(3n) = 2(4n + 3)$$

Now simplify and solve for n:

$$9n = 8n + 6$$
$$n = 6$$

14. **J.** An equilateral triangle divides into two 30-60-90 triangles, whose sides have a ratio of $x : x\sqrt{3} : 2x$. The height of 9 corresponds to the $x\sqrt{3}$, so

$$x\sqrt{3} = 9$$
$$x = \frac{9}{\sqrt{3}}$$

You can simplify this value as I show you in Chapter 5:

$$\frac{9}{\sqrt{3}} = \frac{9}{\sqrt{3}} \cdot \frac{\sqrt{3}}{\sqrt{3}} = \frac{9\sqrt{3}}{3} = 3\sqrt{3}$$

So the base of the triangle is twice this value, which is $6\sqrt{3}$. Plug the lengths of the base and the height into the formula for the area of a triangle:

$$A = \frac{1}{2}bh = \frac{1}{2}(6\sqrt{3})(9) = 27\sqrt{3}$$

15. C. The question gives you two of the three interior angles of the triangle: $x°$ and 100°. The remaining interior angle is supplementary with $y°$, so it's $(180 - y)°$. Thus, you can make the following equation:

$$x + 100 + (180 - y) = 180$$

Solve for y in terms of x:

$$\begin{aligned} x + 100 + 180 - y &= 180 \\ x + 100 - y &= 0 \\ x + 100 &= y \end{aligned}$$

16. G. Write "15% of n is 300" as an equation:

$$0.15n = 300$$

Now solve for n:

$$n = \frac{300}{0.15} = 2{,}000$$

Twenty-two percent of 2,000 is 440.

17. E. Any line perpendicular to $y = \frac{1}{3}x + 9$ has a slope of –3. So you can rule out Choices (A), (B), and (C). Plug this number into the slope-intercept form, along with the x- and y-coordinates for the point (3, 4):

$$\begin{aligned} y &= mx + b \\ 4 &= -3(3) + b \\ 4 &= -9 + b \\ 13 &= b \end{aligned}$$

Now plug the slope $m = -3$ and the y-intercept of 13 into the slope-intercept form to get the formula of the line:

$$y = -3x + 13$$

18. F. To begin, isolate the absolute value on one side of the equation:

$$\begin{aligned} |5m - 11| - 3m &= 9 \\ |5m - 11| &= 9 + 3m \end{aligned}$$

Next, split the equation into two separate equations and remove the absolute value bars:

$$5m - 11 = 9 + 3m \qquad 5m - 11 = -(9 + 3m)$$

Solve both equations for m:

$$\begin{aligned} 5m &= 20 + 3m \\ 2m &= 20 \\ m &= 10 \end{aligned} \qquad \begin{aligned} 5m - 11 &= -9 - 3m \\ 5m &= 2 - 3m \\ 8m &= 2 \\ m &= 0.25 \end{aligned}$$

The product of these two values is $10 \times 0.25 = 2.5$.

19. A. Plug the values $(-3, -7)$ and $(1, 8)$ into the midpoint formula:

$$\text{Midpoint} = \left(\frac{x_1 + x_2}{2}, \frac{y_1 + y_2}{2}\right) = \left(\frac{-3+1}{2}, \frac{-7+8}{2}\right) = \left(-1, \frac{1}{2}\right)$$

20. J. To start, find the values of $f(4)$ and $g(-1)$:

$$f(4) = 4^2 + 9 = 16 + 9 = 25$$
$$g(-1) = 24 + 4(-1) = 24 - 4 = 20$$

Thus:

$$\frac{f(4)}{g(-1)} = \frac{25}{20} = 1.25$$

21. C. The two legs of the triangle are of lengths y and $3y$, and the hypotenuse is of length y. Plug these values into the Pythagorean theorem:

$$a^2 + b^2 = c^2$$
$$y^2 + (3y)^2 = x^2$$

Simplify and solve for x in terms of y:

$$\begin{aligned} y^2 + 9y^2 &= x^2 \\ 10y^2 &= x^2 \\ \sqrt{10y^2} &= x \\ y\sqrt{10} &= x \end{aligned}$$

22. K. The variables v and w are inversely proportional, so for some constant k, the equation $vw = k$ is always true. Thus, when $v = 7$ and $w = 14$:

$$vw = (7)(14) = 98$$

So $k = 98$. When $v = 2$, you can find w like this:

$$\begin{aligned} vw &= 98 \\ 2w &= 98 \\ w &= 49 \end{aligned}$$

23. C. The ratio of girls to boys was 4 to 5, so write the ratio like this:

$$\frac{\text{Girls}}{\text{Boys}} = \frac{4}{5}$$

If you let g equal the number of girls on the trip, you know that the number of boys was $g + 6$. Plug these values into the ratio:

$$\frac{g}{g+6} = \frac{4}{5}$$

Cross-multiply and solve for g:

$$5g = 4(g+6)$$
$$5g = 4g + 24$$
$$g = 24$$

So now you know that 24 girls went on the field trip, and you're ready to find the number of adults. The ratio of adults to girls was 1:4. That is, the number of adults was $\frac{1}{4}$ the number of girls, so you know that 6 adults attended the field trip.

24. H. Line a goes "down 5, over 8," so its slope is $-\frac{5}{8}$. Line b is parallel, so it has the same slope and has a y-intercept of –3. Plug these numbers into the slope-intercept form to get the equation:

$$y = mx + b$$
$$y = -\frac{5}{8}x - 3$$

25. B. The bag contains a total of $7 + 12 + 17 = 36$ socks. Of these, $7 + 17 = 24$ are NOT white. Plug these two numbers (the number of colored socks and the total number of socks) into the formula for probability:

$$\text{Probability} = \frac{\text{Target outcomes}}{\text{Total outcomes}} = \frac{24}{36} = \frac{2}{3}$$

26. H. Remember that $\tan\theta = \frac{O}{A}$, so if $\tan\theta = \frac{4}{3}$, the opposite and adjacent sides of the triangle are in a ratio of 4:3. Thus, the triangle is a 3-4-5 triangle (you can verify this with the Pythagorean theorem), so you can make the following sketch:

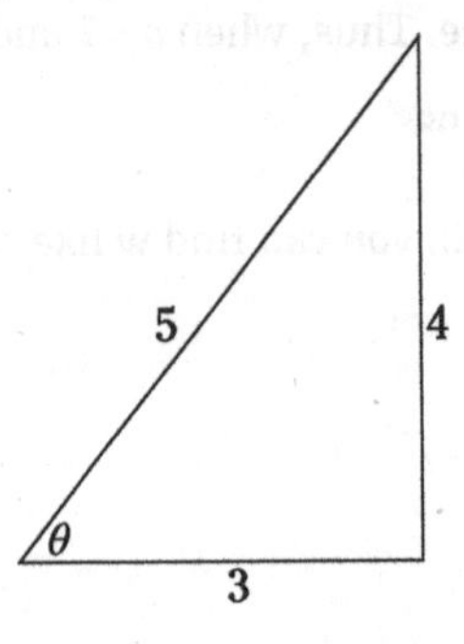

Use this figure and the formula for the sine to answer the question:

$$\sin\theta = \frac{O}{H} = \frac{4}{5}$$

27. E. To begin, notice that the first term of $p^3q^4 + p^4q^5$ contains $(pq)^3$ multiplied by an extra q, and the second term contains $(pq)^4$ multiplied by an extra q. As a result, you can factor those values out of each respective term to simplify:

$$p^3q^4 + p^4q^5 = (pq)^3 q + (pq)^4 q$$

Now you can substitute 3 for pq and simplify:

$$\begin{aligned} &= (3)^3 q + (3)^4 q \\ &= 27q + 81q \\ &= 108q \end{aligned}$$

28. G. The area of the park is 67,500 square feet. If you let w equal the width of the park, the length is $3w$. Plug these numbers into the area formula for a rectangle and solve for w:

$$\begin{aligned} A &= lw \\ 67{,}500 &= (3w)(w) \\ 67.500 &= 3w^2 \\ 22.500 &= w^2 \\ 150 &= w \end{aligned}$$

If the width is 150 feet, the length is $150 \times 3 = 450$ feet. Plug these numbers into the formula for the perimeter of a rectangle:

$$P = 2l + 2w = 2(450) + 2(150) = 900 + 300 = 1{,}200$$

The perimeter of the park is 1,200 feet. Jane ran at 10 feet per second, so she ran for 120 seconds (because $1{,}200 \div 10 = 120$).

29. A. Danielle pays $0.10 per minute, so the function includes 0.1x. However, she gets 200 minutes with her initial $30, so the input x needs to be changed to $x - 200$ to account for this. Therefore, the function includes $0.1(x - 200)$. Additionally, she's charged $30, so the function is:

$$f(x) = 0.1(x - 200) + 30$$

Simplify by distributing and combining to get the final function:

$$\begin{aligned} f(x) &= 0.1x - 20 + 30 \\ f(x) &= 0.1x + 10 \end{aligned}$$

30. H. Use the function you found for Question 29:

$$f(x) = 0.1x + 10$$

Plug in 100 for x and solve for x:

$$\begin{aligned} 100 &= 0.1x + 10 \\ 90 &= 0.1x \\ 900 &= x \end{aligned}$$

31. A. To begin, note that ΔSUR is a right triangle with a leg the length of 9 and a hypotenuse the length of 15, so it's a 9-12-15 version of a 3-4-5 triangle; therefore, $SU = 12$. ΔTUS is a right triangle with a leg the length of 12 and a hypotenuse the length of 13, so it's a 5-12-13 triangle; therefore, $UT = 5$.

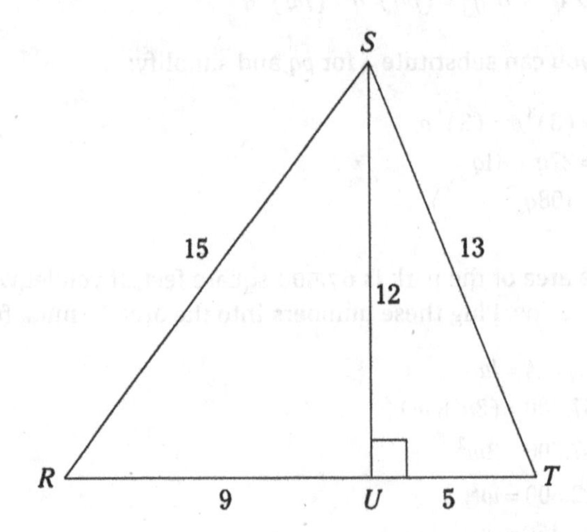

So the base of ΔRST is 14, and its height is 12. Plug these values into the formula for the area of a triangle to get your answer:

$$A = \frac{1}{2}bh = \frac{1}{2}(14)(12) = 84$$

32. G. Let x be the original price of the guitar before tax. Antoine paid $588.60 with a 9% increase on top of the original price. Therefore, Antoine paid 109% of the original price, so you can create this equation:

$$1.09x = 588.60$$

Solve for x:

$$x = \frac{588.60}{1.09} = 540$$

33. C. The x-intercept of a line is the point where it crosses the x-axis — that is, where $y = 0$ — so the second coordinate must equal 0. Therefore, you can rule out Choices (A) and (B). To choose correctly among the three remaining answers, substitute 0 for y in the equation and solve for x:

$$y = 2x - 8$$
$$0 = 2x - 8$$
$$8 = 2x$$
$$4 = x$$

Thus, the x-intercept is (4, 0).

34. G. The determinant of a matrix is a number, not a matrix, so rule out Choices (H), (J), and (K). To determine which of the remaining answers is correct, use the determinant formula $ad - bc$:

$$(3 \times 2) - (-6 \times 1)$$

Simplify:

$$= 6 - (-6) = 6 + 6 = 12$$

35. C. Plug the values $(-12, -7)$ and $(12, 3)$ into the distance formula:

$$\text{Distance} = \sqrt{(x_2 - x_1)^2 + (y_2 - y_1)^2} = \sqrt{(12 - (-12))^2 + (3 - (-7))^2}$$

Now simplify to get the answer:

$$= \sqrt{(24)^2 + (10)^2} = \sqrt{576 + 100} = \sqrt{676} = 26$$

36. J. Put both sides of the equation in terms of powers of 7:

$$\left(\frac{1}{49}\right)^{n+3} = \sqrt{7}$$

$$\left(7^{-2}\right)^{n+3} = 7^{\frac{1}{2}}$$

On the left side, multiply the two exponents:

$$7^{-2(n+3)} = 7^{\frac{1}{2}}$$

The two bases are equal, so the two exponents are equal as well:

$$-2(n+3) = \frac{1}{2}$$

Multiply both sides by 2 to eliminate the fraction, and then simplify and solve for n:

$$-4(n+3) = 1$$

$$-4n - 12 = 1$$

$$-4n = 13$$

$$n = -\frac{13}{4}$$

37. D. Begin by drawing a picture of the ladder and wall:

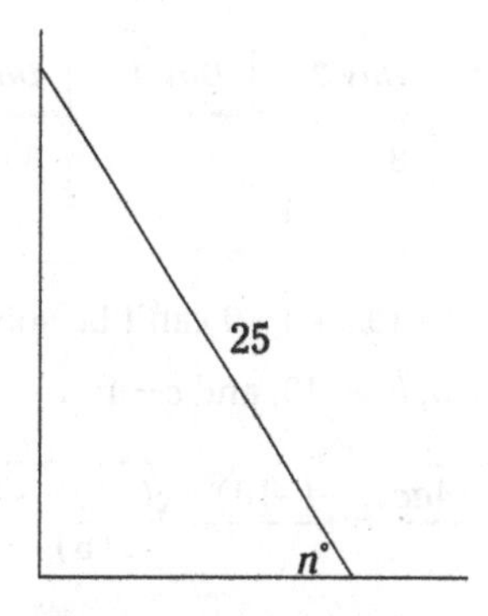

Note that the question asks you to find the distance from the base of the ladder to the wall, which is the adjacent side of this triangle. Begin by using the sine of n, which is the ratio of the opposite side over the hypotenuse:

$$\sin n = \frac{O}{H} = \frac{4}{5}$$

The hypotenuse is 25, so you can set up a proportion to find the length of the opposite side:

$$\frac{O}{25}=\frac{4}{5}$$
$$5O=100$$
$$O=20$$

Now use the Pythagorean theorem to find the length of the adjacent side:

$$20^2+6^2=25^2$$
$$400+b^2=625$$
$$b^2=225$$
$$b=15$$

38. **F.** Begin by plugging the height and the two bases into the formula for the area of a trapezoid:

$$\text{Area}=\frac{b_1+b_2}{2}h$$
$$144=\frac{6x+12x}{2}(4x)$$
$$144=\frac{18x}{2}(4x)$$

Continue simplifying and solve for x:

$$144=9x(4x)$$
$$144=36x^2$$
$$4=x^2$$
$$2=x$$

39. **C.** Ansgar writes for at least 4 hours a day, so in 7 days he writes for at least 28 hours (because $4\times7=28$). On any day that he wrote for 8 hours, he would have written for an additional 4 hours over the minimum. Thus, the week he wrote 46 hours, he wrote for an extra 18 hours (because $46-28=18$). As a result, he could have written for an additional 4 hours on no more than 4 different days. For example, here's one possible schedule:

Day 1	*Day 2*	*Day 3*	*Day 4*	*Day 5*	*Day 6*	*Day 7*	*Total*
8	8	8	8	6	4	4	46

40. **H.** The equation $5x^2-10x+4=0$ can't be solved for x by factoring, so use the quadratic equation, using $a=5$, $b=-10$, and $c=4$:

$$x=\frac{-b\pm\sqrt{b^2-4ac}}{2a}=\frac{-(-10)\pm\sqrt{(-10)^2-4(5)(4)}}{2(5)}$$

Start by simplifying:

$$=\frac{10\pm\sqrt{100-80}}{10}=\frac{10\pm\sqrt{20}}{10}=\frac{10\pm\sqrt{4}\sqrt{5}}{10}=\frac{10\pm2\sqrt{5}}{10}$$

Now split the fraction into two fractions and reduce both separately:

$$=\frac{10}{10}\pm\frac{2\sqrt{5}}{10}=1\pm\frac{\sqrt{5}}{5}$$

41. **A.** Let x be the number, and then translate the words into the following equation:

$$3x+40=2(x+17)$$

Simplify and solve for x:

$$3x+40=2x+34$$
$$x+40=34$$
$$x=-6$$

If you subtract 9 from x and then multiply by 4, the result is:

$$4(-6-9)=4(-15)=-60$$

42. **F.** To solve, you want to multiply each equation by a number so that one of the two variables ends up with the same coefficient in both equations. The easiest way to do so is to multiply every term in $3x+y=-3$ by 4, and then subtract one equation from the other:

$$7x+4y=18$$
$$-\underline{12x+4y=-12}$$
$$-5x\quad=30$$

Next, solve for x:

$$-6=x$$

Finally, plug -6 for x into one of the original equations and solve:

$$3x+y=-3$$
$$3(-6)+y=-3$$
$$-18+y=-3$$
$$y=15$$

Now you know that $x+y=-6+15=9$.

43. **B.** The area of the shaded region is 20% of the whole circle, so $d°$ is 20% of 360°:

$$d=(0.2)(360)=72$$

Use the formula for converting degrees to radians, and plug in 72 for *degrees* and r for *radians:*

$$\frac{180}{\pi}=\frac{\text{degrees}}{\text{radians}}$$
$$\frac{180}{\pi}=\frac{72}{r}$$

Cross-multiply and solve for r:

$$180r=72\pi$$
$$r=\frac{72r}{180}$$
$$r=\frac{2}{5}\pi$$

44. F. Each side of the large square has a length of $x + 2x = 3x$. The area of the large square is 81, so each side of the square is $\sqrt{81} = 9$. So one side of the square is $3x$, which is equal to 9. Thus:

$$3x = 9$$
$$x = 3$$

So each triangle has legs of $x = 3$ and $2x = 6$. Plug these values into the Pythagorean theorem:

$$a^2 + b^2 = c^2$$
$$3^2 + 6^2 = c^2$$

Because c is the side of the shaded square, c^2 is the area of this square. So solve for c^2:

$$9 + 36 = c^2$$
$$45 = c^2$$

Thus, the area of the shaded square is 45.

45. A. The function $g(x)$ is the transformation that moves $f(x) = x^2 + 10x + 2$ one unit up and one unit to the right. To move one unit up, add 1 to the entire function. And to move one unit to the right, substitute $x - 1$ for x in the function. Thus

$$g(x) = f(x-1) + 1$$

Thus, you need to substitute $x - 1$ for x throughout the $f(x)$ and add 1 to $f(x)$:

$$g(x) = (x-1)^2 + 10(x-1) + 2 + 1$$

Now simplify:

$$= (x-1)(x-1) + 10(x-1) + 2 + 1$$
$$= x^2 - 2x + 1 + 10x - 10 + 2 + 1$$
$$= x^2 + 8x - 6$$

46. J. Cross-multiply to get the two fractions out of the equation:

$$\frac{a+b}{10} = \frac{a - 0.1b^2}{a-b}$$
$$(a+b)(a-b) = 10(a - 0.1b^2)$$

FOIL the left side and distribute the right side:

$$a^2 - b^2 = 10a - b^2$$

Add b^2 to both sides of the equation, and then divide by a:

$$a^2 = 10a$$
$$a = 10$$

47. C. The first character must be a letter, so 26 possibilities exist for this character. The last (6th) character must be a digit, so 10 possibilities exist for this one. The remaining 4 characters can be either a letter or a digit, so 36 possibilities exist for each of these. The following chart organizes this information:

1st	*2nd*	*3rd*	*4th*	*5th*	*6th*
26	36	36	36	36	10

Multiply these results:

$$26 \times 36 \times 36 \times 36 \times 36 \times 10 = 436{,}700{,}160$$

This result has 9 digits, so it's between 10^8 (100,000,000) and 10^9 (1,000,000,000).

48. **J.** The formula for a circle of radius r is $(x-h)^2 + (y-k)^2 = r^2$. So in the equation $(x+3)^2 + (y-4)^2 = 49$:

$$r^2 = 49$$
$$r = 7$$

You now plug this value into the formula for the area of a circle to get your answer:

$$A = \pi r^2 = \pi(7)^2 = 49\pi$$

49. **A.** First, use the reciprocal identity $\sec x = \frac{1}{\cos x}$ to substitute for sec x:

$$\sin x \sec x = \frac{\sin x}{\cos x}$$

Now recall the following identity:

$$\frac{\sin x}{\cos x} = \tan x$$

50. **G.** The height of the tank is h and its diameter is $3h$, so its radius is $1.5h$. The volume of the tank is approximately 231.5 cubic meters. Use 3.14 as an approximation of π and plug these values into the formula for a cylinder:

$$V = \pi r^2 h$$
$$231.5 \approx (3.14)(1.5h)^2 h$$

Simplify and solve for h:

$$231.5 \approx (3.14)(2.25h^2)h$$
$$231.5 \approx 7.065h^3$$
$$32.767 \approx h^3$$
$$3.2 \approx h$$

Thus, the height of the tank is closest to 3 meters.

51. **D.** Let p, q, and r equal the amounts that Paulette, Quentin, and Rosie gave, respectively. Then you can set up the following three equations:

$$p = q + r \qquad 3q = p + 40 \qquad 2(r - 20) = p$$

Simplify the third equation

$$p = q + r \qquad 3q = p + 40 \qquad 2r - 40 = p$$

Substitute $q + r$ for p into the second and third equations:

$$3q = q + r + 40 \qquad 2r - 40 = q + r$$

Simplify both equations:

$$2q = r + 40 \qquad r - 40 = q$$

Now substitute $r - 40$ for q into the equation $2q = r + 40$ and solve for r:

$$2(r - 40) = r + 40$$
$$2r - 80 = r + 40$$
$$r - 80 = 40$$
$$r = 120$$

Substitute 120 for r in the equation $2r - 40 = p$ and solve for p:

$$2(120) - 40 = p$$
$$240 - 40 = p$$
$$200 = p$$

52. **G.** Zach registered 12 clients, Yolanda registered 16, Xavier registered 19, Wanda registered 13, and Victoria registered 20. Add these together to find the total number of clients registered:

$$12 + 16 + 19 + 13 + 20 = 80$$

So Yolanda registered:

$$\frac{16}{80} = \frac{1}{5} = 20\%$$

53. **B.** If Victoria doubles her registration to 40 and the others keep the same numbers, the sum of registrations will be:

$$12 + 16 + 19 + 13 + 40 = 100$$

Thus, Victoria's registration will be:

$$\frac{40}{100} = 40\%$$

54. **H.** The octagon is regular, so the shaded region is a right triangle. It has a base of 1, so you need to know its height to find its area. Draw a few lines as follows to help find the height of the triangle:

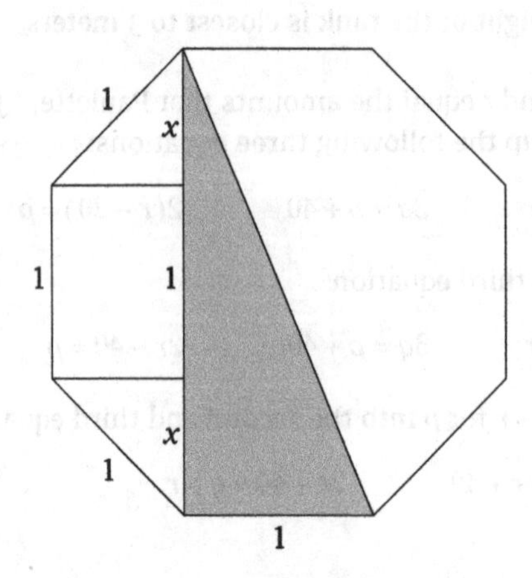

If you let x equal the unknown length, the height of the shaded region is $2x+1$. To find the value of x, notice that the length x is the leg of a 45-45-90 triangle with a hypotenuse of 1. The ratio of a leg of this triangle to its hypotenuse is $1:\sqrt{2}$ Thus:

$$\frac{1}{\sqrt{2}}=\frac{x}{1}$$

Therefore:

$$x=\frac{1}{\sqrt{2}}=\frac{\sqrt{2}}{2}$$

So

$$2x+1=\sqrt{2}+1$$

Plug this value as the height into the formula for the area of a triangle, with a base of 1:

$$A=\frac{1}{2}bh=\frac{1}{2}(1)\left(\sqrt{2}+1\right)=\frac{\sqrt{2}+1}{2}$$

55. B. Begin by multiplying all three terms by a common denominator of abc to get rid of the fractions:

$$\frac{a}{c}-\frac{a}{b}=\frac{b-c}{a}$$

$$(abc)\frac{a}{c}-(abc)\frac{a}{b}=(abc)\frac{b-c}{a}$$

When the denominators are canceled out, the result is the following equation:

$$a^2b-a^2c=bc(b-c)$$

Factor out a^2 on the left side of the equation:

$$a^2(b-c)=bc(b-c)$$

Now divide both sides of the equation by $(b-c)$ and cancel:

$$\frac{a^2(b-c)}{(b-c)}=\frac{bc(b-c)}{(b-c)}$$

$$a^2=bc$$

Finally, take the square root of both sides:

$$a=\sqrt{bc}$$

56. K. If you let x be the speed at which Angela ran, you can let $x-4$ be the speed at which Angela and Kathleen walked. The distance in each direction was 2 miles, and the total time was 45 minutes, which is $\frac{3}{4}$ of an hour. Place all of this information into a rate-time-distance chart:

	Rate	*Time*	*Distance*
Running	x		2
Walking	$x-4$		2
Total		$\frac{3}{4}$	

Rate × Time = Distance, so, in your chart, calculate Time = Distance ÷ Rate:

	Rate	*Time*	*Distance*
Running	x	$\frac{2}{x}$	2
Walking	$x-4$	$\frac{2}{x-4}$	2
Total		$\frac{3}{4}$	

Adding the Time column, set up the following equation:

$$\frac{2}{x}+\frac{2}{x-4}=\frac{3}{4}$$

Use $x(x-4)$ as a common denominator on the left side of the equation to add the two fractions and simplify:

$$\frac{2(x-4)+2x}{x(x-4)}=\frac{3}{4}$$
$$\frac{2x-8+2x}{x^2-4x}=\frac{3}{4}$$
$$\frac{4x-8}{x^2-4x}=\frac{3}{4}$$

Now cross-multiply and simplify again:

$$4(4x-8)=3\left(x^2-4x\right)$$
$$16x-32=3x^2-12x$$
$$0=3x^2-28x+32$$

Solve the resulting quadratic equation for x using either factoring or the quadratic formula (I use factoring):

$$(3x-4)(x-8)=0$$
$$3x-4=0 \qquad x-8=0$$

The first equation solves for x as a number that's less than $4\left(x=\frac{4}{3}\right)$, which isn't correct in the context of the question because their speed on the way back would be negative. The second equation solves as $x=8$, so the correct answer is Choice (K).

57. **C.** The function $f(x)$ is symmetrical, so if you reflect it horizontally across the y-axis and then shift it 6 units to the right, it returns to where it started. To reflect it horizontally, change x to $-x$:

$$f(x)$$

Now, to move this function 6 units to the right, change x in this new function to $x - 6$ and simplify:

$$f(-(x-6)) = f(-x+6) = f(6-x)$$

58. **G.** Convert the log into an exponent:

$$\log_3 n = \frac{1}{2} \qquad \text{means} \qquad 9^{\frac{1}{2}} = n$$

Simplify, keeping in mind that $9^{\frac{1}{2}} = \sqrt{9}$:

$$\sqrt{9} = n$$
$$3 = n$$

Thus, $\sqrt{n} = \sqrt{3}$

59. **D.** The conjugate of $2+7i$ is $2-7i$, so you're looking for the value of $(2+7i)(2-7i)$. Begin by FOILing to remove the parentheses:

$$(2+7i)(2-7i) = 4 - 14i + 14i - 49i^2 = 4 - 49i^2$$

Now, because $i = \sqrt{-1}$, you can substitute -1 for i^2:

$$4 - 49(-1) = 4 + 49 = 53$$

60. **K.** The water level decreases and then rises again, which rules out Choices (F) and (H). The pool is full to capacity at the beginning of the shift, so it can't have more water at the end of the shift; therefore, Choice (J) is also wrong. Finally, the pool drains in 2 hours but takes 7 hours to fill, so the downward slope at the beginning of the shift is greater than the upward slope at the end; as a result, you can rule out Choice (G), leaving Choice (K) as the correct answer.

Answer Key for Practice Test 1

1.	D	21.	C	41.	A
2.	J	22.	K	42.	F
3.	A	23.	C	43.	B
4.	H	24.	H	44.	F
5.	D	25.	B	45.	A
6.	G	26.	H	46.	J
7.	C	27.	E	47.	C
8.	G	28.	G	48.	J
9.	D	29.	A	49.	A
10.	H	30.	H	50.	G
11.	D	31.	A	51.	D
12.	J	32.	G	52.	G
13.	A	33.	C	53.	B
14.	J	34.	G	54.	H
15.	C	35.	C	55.	B
16.	G	36.	J	56.	K
17.	E	37.	D	57.	C
18.	F	38.	F	58.	G
19.	A	39.	C	59.	D
20.	J	40.	H	60.	K

Chapter 16
Practice Test 2

I hope that you completed Practice Test 1 with flying colors and that this one is simply for extra preparation. If not, best of luck here. Make your way through this practice test to find out where you stand and further reinforce all that information you've crammed into your head.

In order to best simulate real exam conditions, I recommend you do the following:

1. **Sit where you won't be interrupted or tempted to pick up the TV remote or your phone.**
2. **Use the answer sheet provided to practice filling in the dots.**
3. **Set your timer for the time limits indicated at the beginning of the test.**
4. **Check your work for this test only; don't look at more than one test at a time.**
5. **Avoid taking breaks during the test.**

TIP

When you finish this practice test, turn to Chapter 17, where you find detailed explanations of the answers as well as an abbreviated answer key. I recommend that you go through the answer explanations to all the questions, not just the ones that you missed, because you'll find lots of good info that may help you later on.

Mathematics Test Answer Sheet

1. Ⓐ Ⓑ Ⓒ Ⓓ Ⓔ
2. Ⓕ Ⓖ Ⓗ Ⓙ Ⓚ
3. Ⓐ Ⓑ Ⓒ Ⓓ Ⓔ
4. Ⓕ Ⓖ Ⓗ Ⓙ Ⓚ
5. Ⓐ Ⓑ Ⓒ Ⓓ Ⓔ
6. Ⓕ Ⓖ Ⓗ Ⓙ Ⓚ
7. Ⓐ Ⓑ Ⓒ Ⓓ Ⓔ
8. Ⓕ Ⓖ Ⓗ Ⓙ Ⓚ
9. Ⓐ Ⓑ Ⓒ Ⓓ Ⓔ
10. Ⓕ Ⓖ Ⓗ Ⓙ Ⓚ
11. Ⓐ Ⓑ Ⓒ Ⓓ Ⓔ
12. Ⓕ Ⓖ Ⓗ Ⓙ Ⓚ
13. Ⓐ Ⓑ Ⓒ Ⓓ Ⓔ
14. Ⓕ Ⓖ Ⓗ Ⓙ Ⓚ
15. Ⓐ Ⓑ Ⓒ Ⓓ Ⓔ
16. Ⓕ Ⓖ Ⓗ Ⓙ Ⓚ
17. Ⓐ Ⓑ Ⓒ Ⓓ Ⓔ
18. Ⓕ Ⓖ Ⓗ Ⓙ Ⓚ
19. Ⓐ Ⓑ Ⓒ Ⓓ Ⓔ
20. Ⓕ Ⓖ Ⓗ Ⓙ Ⓚ
21. Ⓐ Ⓑ Ⓒ Ⓓ Ⓔ
22. Ⓕ Ⓖ Ⓗ Ⓙ Ⓚ
23. Ⓐ Ⓑ Ⓒ Ⓓ Ⓔ
24. Ⓕ Ⓖ Ⓗ Ⓙ Ⓚ
25. Ⓐ Ⓑ Ⓒ Ⓓ Ⓔ
26. Ⓕ Ⓖ Ⓗ Ⓙ Ⓚ
27. Ⓐ Ⓑ Ⓒ Ⓓ Ⓔ
28. Ⓕ Ⓖ Ⓗ Ⓙ Ⓚ
29. Ⓐ Ⓑ Ⓒ Ⓓ Ⓔ
30. Ⓕ Ⓖ Ⓗ Ⓙ Ⓚ
31. Ⓐ Ⓑ Ⓒ Ⓓ Ⓔ
32. Ⓕ Ⓖ Ⓗ Ⓙ Ⓚ
33. Ⓐ Ⓑ Ⓒ Ⓓ Ⓔ
34. Ⓕ Ⓖ Ⓗ Ⓙ Ⓚ
35. Ⓐ Ⓑ Ⓒ Ⓓ Ⓔ
36. Ⓕ Ⓖ Ⓗ Ⓙ Ⓚ
37. Ⓐ Ⓑ Ⓒ Ⓓ Ⓔ
38. Ⓕ Ⓖ Ⓗ Ⓙ Ⓚ
39. Ⓐ Ⓑ Ⓒ Ⓓ Ⓔ
40. Ⓕ Ⓖ Ⓗ Ⓙ Ⓚ
41. Ⓐ Ⓑ Ⓒ Ⓓ Ⓔ
42. Ⓕ Ⓖ Ⓗ Ⓙ Ⓚ
43. Ⓐ Ⓑ Ⓒ Ⓓ Ⓔ
44. Ⓕ Ⓖ Ⓗ Ⓙ Ⓚ
45. Ⓐ Ⓑ Ⓒ Ⓓ Ⓔ
46. Ⓕ Ⓖ Ⓗ Ⓙ Ⓚ
47. Ⓐ Ⓑ Ⓒ Ⓓ Ⓔ
48. Ⓕ Ⓖ Ⓗ Ⓙ Ⓚ
49. Ⓐ Ⓑ Ⓒ Ⓓ Ⓔ
50. Ⓕ Ⓖ Ⓗ Ⓙ Ⓚ
51. Ⓐ Ⓑ Ⓒ Ⓓ Ⓔ
52. Ⓕ Ⓖ Ⓗ Ⓙ Ⓚ
53. Ⓐ Ⓑ Ⓒ Ⓓ Ⓔ
54. Ⓕ Ⓖ Ⓗ Ⓙ Ⓚ
55. Ⓐ Ⓑ Ⓒ Ⓓ Ⓔ
56. Ⓕ Ⓖ Ⓗ Ⓙ Ⓚ
57. Ⓐ Ⓑ Ⓒ Ⓓ Ⓔ
58. Ⓕ Ⓖ Ⓗ Ⓙ Ⓚ
59. Ⓐ Ⓑ Ⓒ Ⓓ Ⓔ
60. Ⓕ Ⓖ Ⓗ Ⓙ Ⓚ

Mathematics Test

TIME: 60 minutes for 60 questions

DIRECTIONS: Each question has five answer choices. Choose the best answer for each question, and then shade in the corresponding oval on your answer sheet.

1. At a convenience store, two candy bars and two bags of potato chips cost $4.00, and three candy bars and two bags of potato chips cost $4.75. What is the price of one bag of potato chips?

(A) $0.50

(B) $0.75

(C) $1.00

(D) $1.25

(E) $1.50

2. Which of the following is the greatest common factor of 60 and 64?

(F) 2

(G) 4

(H) 6

(J) 8

(K) 10

3. A teacher takes 5 children on a field trip to a geological museum. As a souvenir, the group receives a bag of stones. When the teacher divides the stones evenly among the children, each child receives exactly 24. If 6 children had been present, how many stones would each child have received?

(A) 15

(B) 16

(C) 18

(D) 20

(E) 21

4. In the following figure, if the perimeter of the square is 20, what is the area of the shaded region?

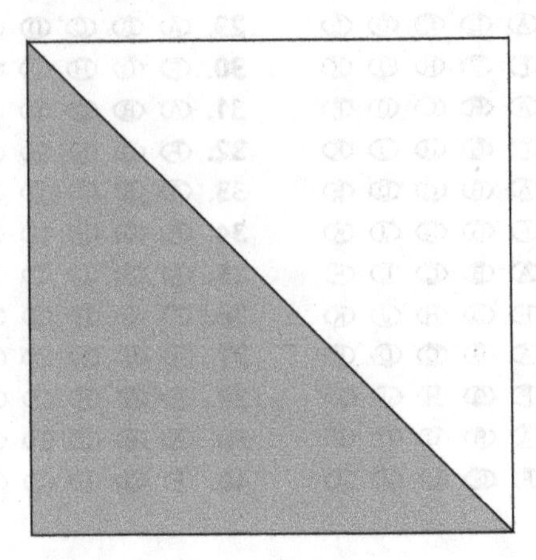

(F) 10

(G) 12

(H) 12.5

(J) 20.5

(K) 25

5. At a banquet of 36 people, every person had a choice among beef stroganoff, chicken divan, and linguini primavera. If 25% chose beef stroganoff and 17 people chose chicken divan, how many people chose linguini primavera?

(A) 7

(B) 8

(C) 9

(D) 10

(E) 11

6. What is the value of $4v\left(w^2-3vw\right)$ given that $v=-1$ and $w=4$

(F) –84

(G) –92

(H) –98

(J) –104

(K) –112

7. Maeve took 10 minutes to walk around a rectangular field. The length of the field is 4 times its width. How long would it take Maeve to walk the width of the field?

(A) 1 minute

(B) 2 minutes

(C) 3 minutes

(D) 4 minutes

(E) 5 minutes

8. What is the slope of the line in the following graph?

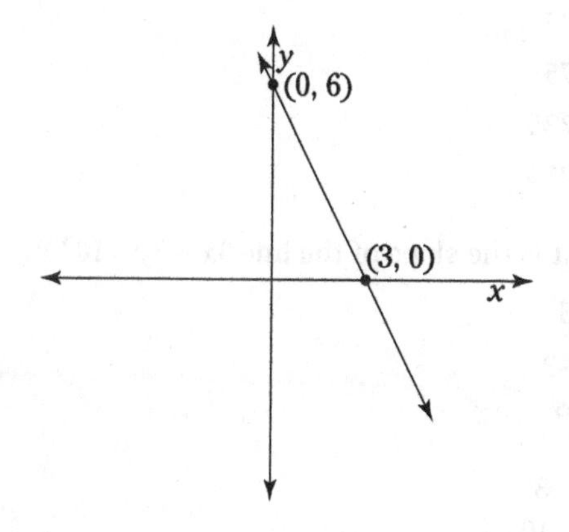

(F) 2

(G) –2

(H) $-\frac{1}{2}$

(J) $\frac{4}{3}$

(K) $-\frac{4}{3}$

9. Two values of v satisfy the equation $|2v-13|+6=9$. What is the sum of these values?

(A) 4

(B) 7

(C) 9

(D) 13

(E) 17

10. Damien played golf on each of the four days of his vacation. His scores on the first three days were 93, 92, and 89, and his average for the four days was 90. What was his score on the fourth day?

(F) 84

(G) 85

(H) 86

(J) 87

(K) 88

11. Which of the following inequalities is equivalent to $7-3p\geq\frac{5}{2}$?

(A) $p\geq\frac{3}{2}$

(B) $p\leq\frac{3}{2}$

(C) $p\geq-\frac{1}{2}$

(D) $p\geq-\frac{3}{2}$

(E) $p\leq-\frac{3}{2}$

12. In the following figure, line M and line N are parallel. Which of the answer choices does NOT necessarily add up to 180°?

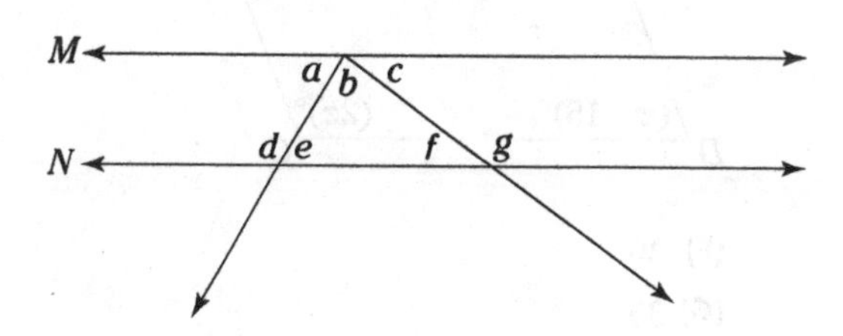

(F) $\angle a+\angle b+\angle f$

(G) $\angle a+\angle d$

(H) $\angle b+\angle e+\angle f$

(J) $\angle d+\angle e$

(K) $\angle e+\angle g$

13. On an xy-graph, what is the length of a line segment drawn from $(-3,7)$ to $(6,-5)$?

(A) 15

(B) 16

(C) 17

(D) 18

(E) 20

GO ON TO NEXT PAGE

14. Which of the following is NOT a factor of $4x^2y^4 - 12x^3y^2 - 8xy^3$?

(F) $2x^2$

(G) $2xy^2$

(H) $-2y$

(J) $4y^2$

(K) $4xy$

15. Of the 126 students who applied for a full scholarship at Oxbow College, 9 received one. What is the ratio of students who received a scholarship to those who didn't?

(A) 1 to 10

(B) 1 to 11

(C) 1 to 12

(D) 1 to 13

(E) 1 to 14

16. In the following figure, each of the four angles in the parallelogram is as shown. What is the value of y?

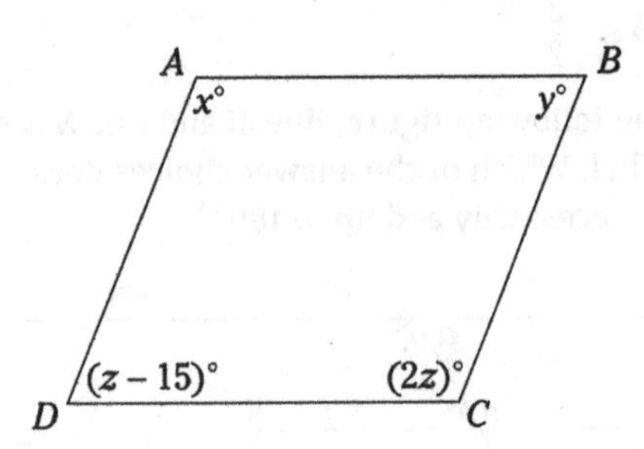

(F) 30

(G) 35

(H) 40

(J) 45

(K) 50

17. If $2x - y = 32$ and $5x + 3y = 14$, then $xy =$

(A) 35

(B) –40

(C) 75

(D) 100

(E) –120

18. What is the value of a in terms b if $\frac{a}{b} + \frac{a+2}{3b} = \frac{1}{4}$?

(F) $\frac{b+4}{2}$

(G) $\frac{b-4}{2}$

(H) $\frac{3b-2}{4}$

(J) $\frac{3b+8}{16}$

(K) $\frac{3b-8}{16}$

19. If p percent of 250 is 75, what is 75% of p?

(A) 22.5

(B) 25

(C) 75

(D) 225

(E) 250

20. What is the slope of the line $9x - 3y = 10$?

(F) 3

(G) –3

(H) 9

(J) $-\frac{1}{3}$

(K) $-\frac{10}{3}$

21. In the following figure, F is the midpoint of $\overline{GH}$. Which of the following are the coordinates of F?

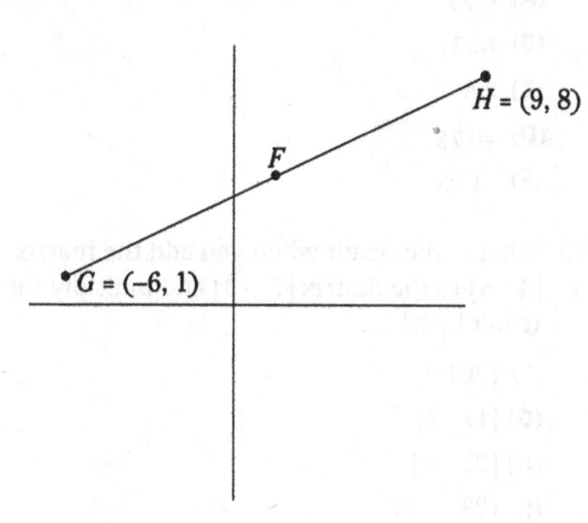

(A) (1, 4)

(B) (2, 5)

(C) $\left(\frac{3}{2}, 4\right)$

(D) $\left(\frac{3}{2}, \frac{9}{2}\right)$

(E) $\left(2, \frac{9}{2}\right)$

22. If $|6 - 4n| > 1$, which of the following must be true?

(F) $\frac{5}{4} < n < \frac{7}{4}$

(G) $-\frac{5}{4} < n < \frac{7}{4}$

(H) $-\frac{7}{4} < n < \frac{5}{4}$

(J) $n < \frac{5}{4}$ or $n > \frac{7}{4}$

(K) $n < -\frac{5}{4}$ or $n > \frac{7}{4}$

23. In the following figure, $\sin\theta =$

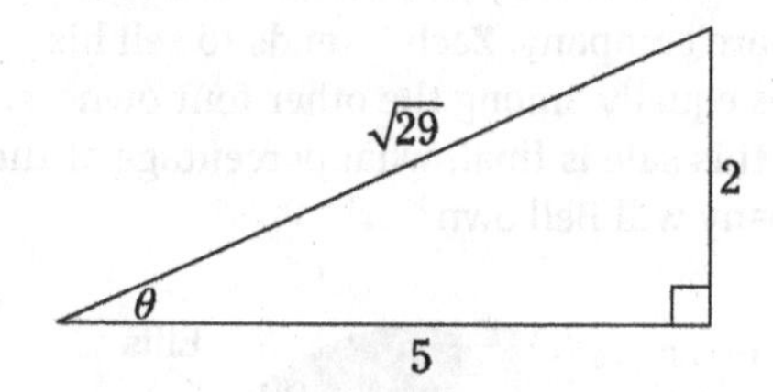

(A) $\frac{2}{5}$

(B) $\frac{5}{2}$

(C) $\frac{\sqrt{29}}{2}$

(D) $\frac{\sqrt{29}}{5}$

(E) $\frac{2\sqrt{29}}{29}$

24. The formula for the volume of a sphere is $V = \frac{4}{3}\pi r^3$, and the formula for the surface area of a sphere is $A = 4\pi r^2$. If a sphere has a surface area of 36π, what is its volume?

(F) 27π

(G) 36π

(H) 54π

(J) 72π

(K) 108π

25. What is a possible value of x if $\sqrt{2x+3} - 1 = x$?

(A) $\sqrt{2}$

(B) $\sqrt{3}$

(C) $\sqrt{5}$

(D) $\sqrt{7}$

(E) $\sqrt{11}$

26. If m and n are both positive factors of 75, which of the following could equal $m + n$?

(F) 12

(G) 18

(H) 22

(J) 36

(K) 54

GO ON TO NEXT PAGE

27. The following graph shows the percentage of shares controlled by five owners of a startup software company. Zach intends to sell his shares equally among the other four owners. After this sale is final, what percentage of the company will Bell own?

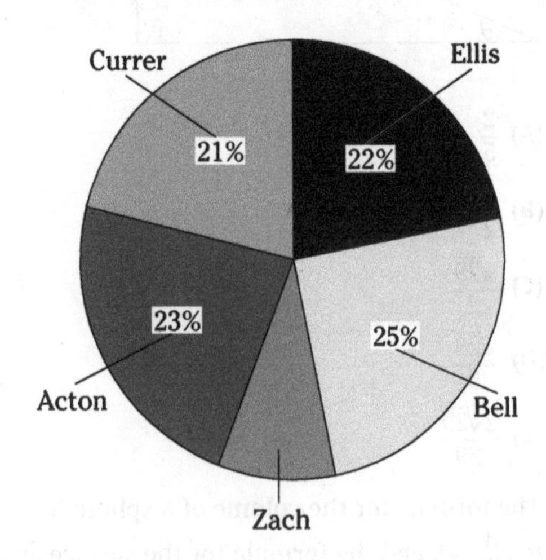

(A) 25.25%

(B) 27.25%

(C) 27.5%

(D) 39.9%

(E) 34%

28. A clown at an amusement park makes animal shapes from twisted balloons. The clown sells each animal based on the number of balloons it requires, according to the following chart:

Number of balloons:

	1	2	3	4	5	6	7
Price:	\$4.00	\$4.50	\$5.00	\$5.50	\$6.00	\$6.50	\$7.00

Which of the following functions allows you to correctly calculate the dollar cost of a balloon animal that contains x balloons?

(F) $f(x) = x$

(G) $f(x) = x + 4$

(H) $f(x) = 4x + 0.5$

(J) $f(x) = 0.5x + 4$

(K) $f(x) = 0.5x + 3.5$

29. What is the sum of the two values of x that satisfy the equation $4x^2 - 3x - 1 = 0$?

(A) 0.75

(B) 1.25

(C) 2.5

(D) –0.75

(E) –1.25

30. What is the result when you add the matrix $[4 \quad 5]$ to the matrix $[7 \quad -3]$ and multiply the result by 2?

(F) $[26]$

(G) $[11 \quad 2]$

(H) $[22 \quad 4]$

(J) $[28 \quad -15]$

(K) $[54 \quad -30]$

31. If the equation for the line shown in the following graph is $y = \frac{1}{3}x + 3$, what is the value of kn?

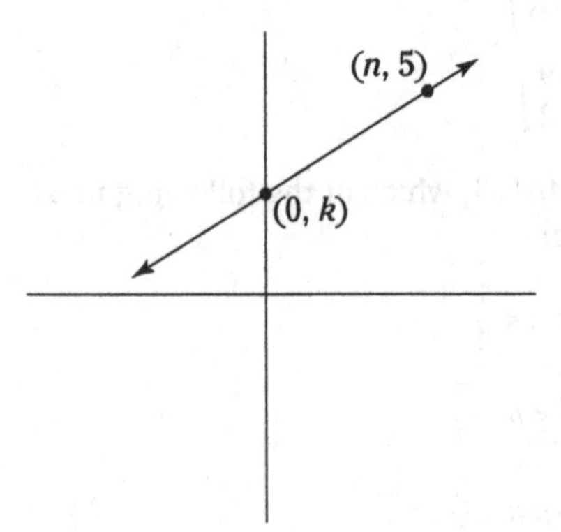

(A) 9

(B) 12

(C) 15

(D) 18

(E) 24

32. Alfred and Rani both picked different two-digit numbers. If you multiply Alfred's number by 5 and double Rani's number, the sum is 300. If you double Alfred's number and multiply Rani's number by 3, the sum of the two numbers is 252. What is the sum of their two numbers?

(F) 96

(G) 112

(H) 128

(J) 144

(K) 150

33. What is the equation of a line that intersects the origin and is perpendicular to $2x - 4y = 13$?

(A) $y = -2$

(B) $y = 2x$

(C) $y = -2x$

(D) $y = \frac{1}{2}x$

(E) $y = \frac{1}{2}x - \frac{13}{4}$

34. This semester, Gerry scored an average of 93 on his five history exams. He got the same score on his first two exams, and then he got a 94, an 85, and a 90 on the remaining exams. What score did he receive on his first two exams?

(F) 95

(G) 96

(H) 97

(J) 98

(K) 99

35. What is the value of x if $\frac{2x + 3y - 19}{y + 5} = 3$?

(A) 17

(B) −17

(C) 29

(D) −29

(E) Cannot be determined from the information given.

36. A scientist performs an experiment in which they measure four values two times each, with the following results:

	w	x	y	z
First measurement	0.2	6	0.5	10
Second measurement	0.6	3	1	30

Which of the following conclusions does the experiment provide evidence for?

(F) w and y are directly proportional.

(G) w and z are inversely proportional.

(H) x and y are inversely proportional.

(J) x and z are inversely proportional.

(K) y and z are directly proportional.

GO ON TO NEXT PAGE

37. Which of the following is the amplitude of the function shown in the following graph?

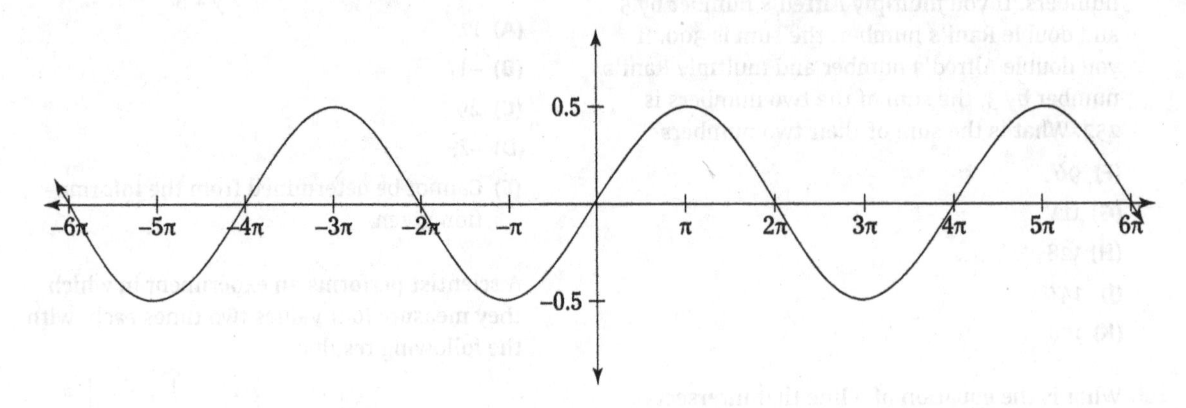

(A) 0.5

(B) 1

(C) π

(D) 2π

(E) 4π

38. In the following figure, O is the center of the circle, $\overline{AB}$ is tangent to the circle at D, and $\overline{BC}$ is tangent to the circle at E. Which of the following must be true?

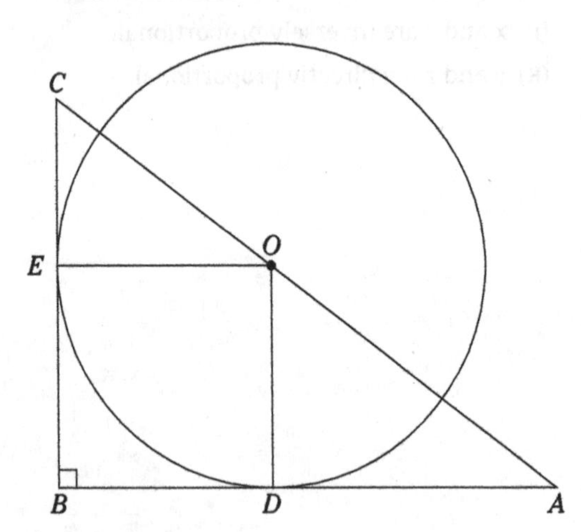

(F) $\overline{OC} \cong \overline{OD}$

(G) ΔABC is isosceles

(H) $\Delta ABC \cong \Delta OEC$

(J) $\Delta OEC \cong \Delta ADO$

(K) $DBEO$ is a square

39. If $g(x)$ is a transformation that moves $f(x)$ three units to the right and then reflects it across the x-axis, then $g(x) =$

(A) $f(-x)+3$

(B) $f(-x)-3$

(C) $f(x)+3$

(D) $-f(x+3)$

(E) $-f(x-3)$

40. The following figure shows the graph of an equation $y = ax^2 + bx + c$. Which of the answer choices CANNOT be true?

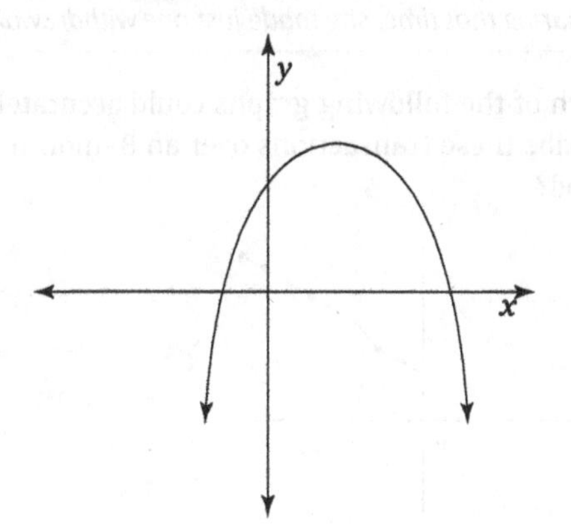

(F) $a < b$

(G) $a > b$

(H) $a < c$

(J) $b < c$

(K) $b > c$

41. Which of the following is the domain of the function $f(x) = \dfrac{3-x}{\sqrt{x^2-9}}$?

(A) $-3 > x > 3$

(B) $-3 \le x \le 3$

(C) $-3 \le x < 3$

(D) $x < -3$ or $x < 3$

(E) $x \le -3$ or $x \ge 3$

42. If a number sequence begins 1, 3, 4, 6, 7, 9, 10, 12 . . ., which of the following numbers does NOT appear in the sequence?

(F) 34

(G) 43

(H) 57

(J) 65

(K) 72

43. A two-digit number from 10 to 99, inclusive, is chosen at random. What is the probability that this number is divisible by 5?

(A) $\frac{1}{5}$

(B) $\frac{2}{9}$

(C) $\frac{19}{90}$

(D) $\frac{18}{91}$

(E) $\frac{19}{91}$

44. What is the area of the shaded region in the following figure?

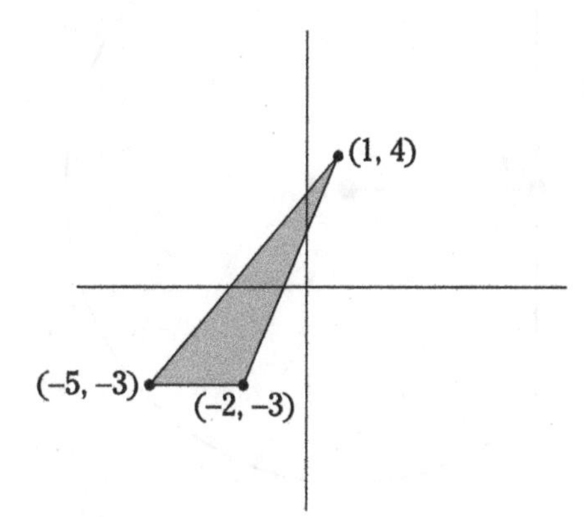

(F) 10.5

(G) 12.5

(H) 18

(J) 21

(K) 24

45. An isosceles triangle contains three angles that measure 40°, x°, and y°. Which of the following CANNOT be true?

(A) $x = y$

(B) $x = 50$

(C) $x - y = 60$

(D) $x = 70$

(E) $x = 100$

GO ON TO NEXT PAGE

46. In the complex numbers, where $i^2 = -1$, what is the value of $5 + 6i$ multiplied by $3 - 2i$?

(F) 27

(G) $27i$

(H) $27 + 8i$

(J) $15 + 8i$

(K) $15 - 8i$

47. In the following figure, $\overline{KL}$ is a chord of the circle centered at O, with $\overline{KL} \perp \overline{MO}$. If $KL = 12$ and $MO = 6$, what is the area of the circle?

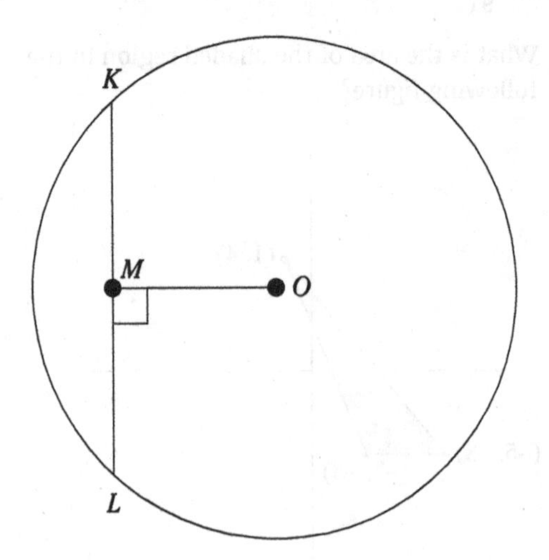

(A) 36π

(B) 48π

(C) 72π

(D) 96π

(E) 144π

Use the following information to answer Questions 48 and 49: Eldridge opened a savings account with an initial balance of $1,000. After that, every month for 8 months, she made one deposit to the account, always for the same amount. During that time, she made just one withdrawal.

48. Which of the following graphs could accurately describe these transactions over an 8-month period?

(F)

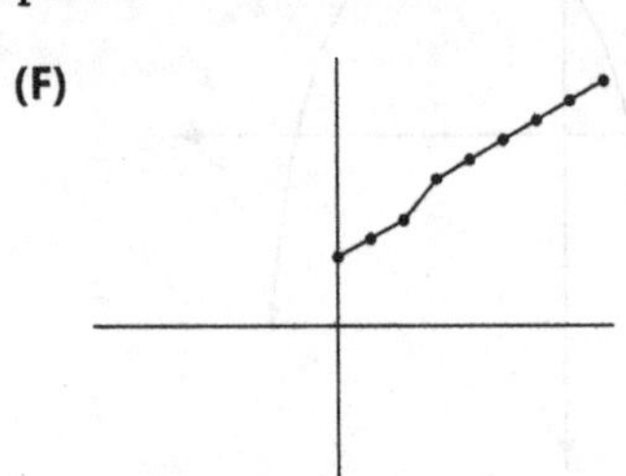

(G)

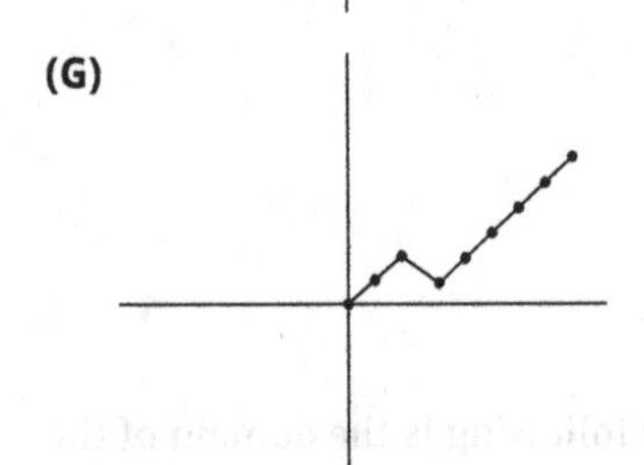

(H)

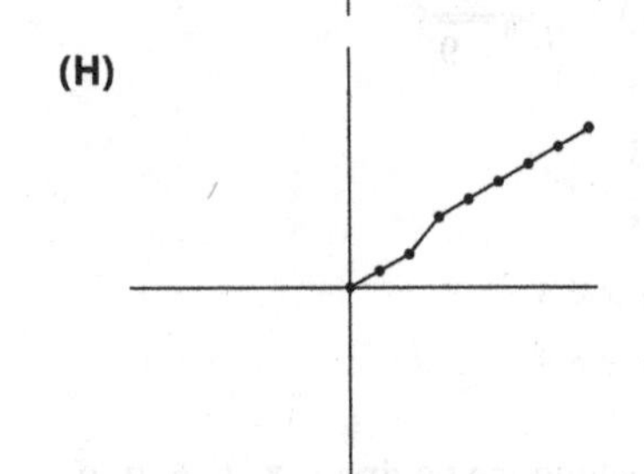

(J)

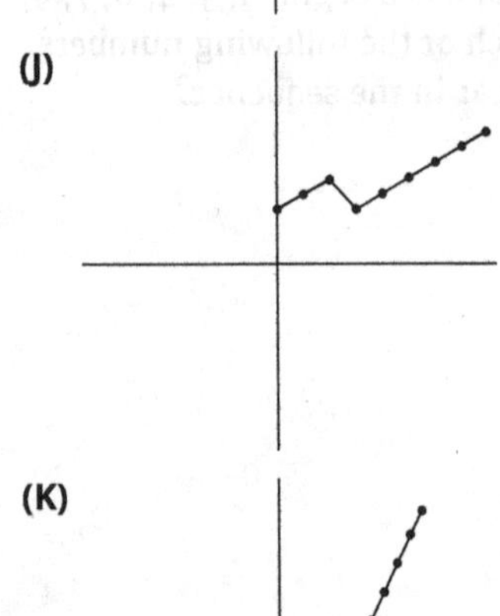

(K)

49. If each of Eldridge's 8 deposits was in the amount of \$200 and her withdrawal was \$350, how much money did her account have at the end of 8 months?

(A) \$850

(B) \$1,550

(C) \$2,250

(D) \$2,350

(E) \$2,950

50. Martha picked out a pair of shoes she wanted and brought them to the front of the store to pay. The cashier told her that the shoes were on sale for 30% off the original price. She rang up the sale price plus 5% sales tax, so Martha ended up paying \$58.80 for the shoes. What was the original price for the shoes?

(F) \$70

(G) \$75

(H) \$80

(J) \$85

(K) \$90

51. Which of the following is equivalent $\frac{\tan n \csc n}{\sin n \sec n}$?

(A) 1

(B) $\sin n$

(C) $\cos n$

(D) $\cot n$

(E) $\csc n$

52. In the following figure, the large quadrilateral is a square. What is the area of the shaded region in terms of x and y?

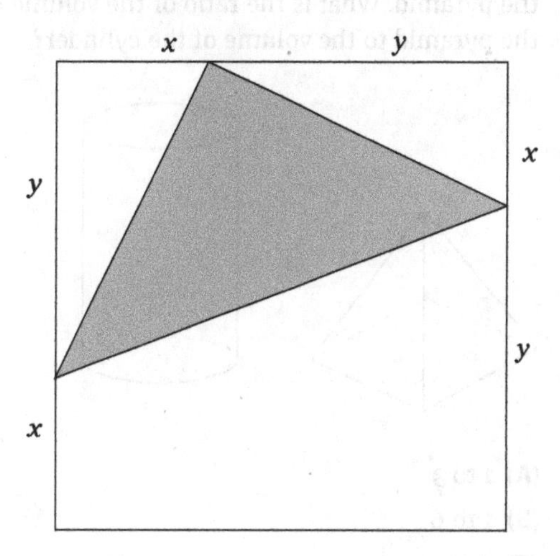

(F) $\frac{x^2 + y^2}{2}$

(G) $\frac{(x + y)^2}{2}$

(H) $\frac{\sqrt{x^2 + y^2}}{2}$

(J) $\frac{x^2 + y^2}{4}$

(K) $\frac{(x + y)^2}{4}$

53. If $\frac{f}{g} = \frac{1}{4}$ and $\frac{g}{h} = \frac{2}{5}$, what is the ratio of $f:h$?

(A) 1:6

(B) 1:8

(C) 1:10

(D) 1:12

(E) 1:20

54. If $49^{3y} = \sqrt{7^{y+1}}$, then $y =$

(F) $\frac{1}{2}$

(G) $\frac{1}{3}$

(H) $\frac{1}{5}$

(J) $\frac{1}{7}$

(K) $\frac{1}{11}$

GO ON TO NEXT PAGE

55. In the following figure, the base of the pyramid has the same area as the base of the cylinder, and the cylinder is twice the height of the pyramid. What is the ratio of the volume of the pyramid to the volume of the cylinder?

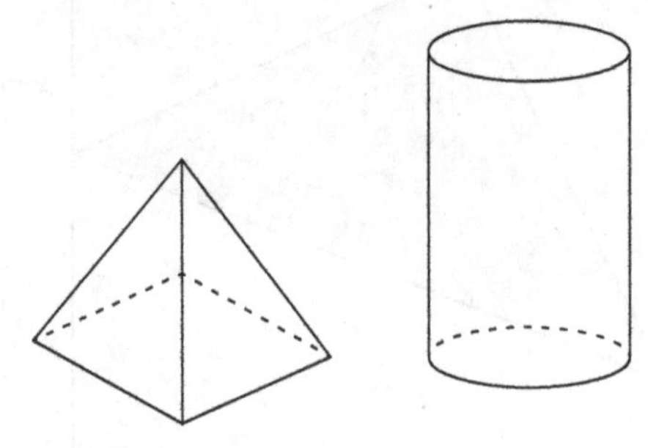

(A) 1 to 3

(B) 1 to 6

(C) 2 to 3

(D) 3 to 2

(E) 6 to 1

56. Which of the following is a possible solution for x in terms of k for the equation $x = \frac{2k}{x+2}$?

(F) $\sqrt{2k}$

(G) $\sqrt{-2k}$

(H) $1 - \sqrt{1+2k}$

(J) $\sqrt{1+2k} + 1$

(K) $\sqrt{1+2k} - 1$

57. If $\frac{a^2 + 2ab + b^2}{a^2 - b^2} = 2a + 2b$, what is the value of $a - b$?

(A) 1

(B) −1

(C) 2

(D) $\frac{1}{2}$

(E) $-\frac{1}{2}$

58. Which of the following functions has a range of $f(x) \geq 4$?

(F) $f(x) = |x+4|$

(G) $f(x) = |x-4|$

(H) $f(x) = |x+4| - 4$

(J) $f(x) = |x-4| + 4$

(K) $f(x) = |x-4| - 4$

59. If $\log_a \sqrt{b} = \frac{1}{4}$ then $a =$

(A) $\sqrt{b}$

(B) b

(C) b^2

(D) b^4

(E) b^8

60. The following figure shows a regular octagon inscribed in a circle. The arc length from A to B is 6π. What is the area of the shaded region of the circle?

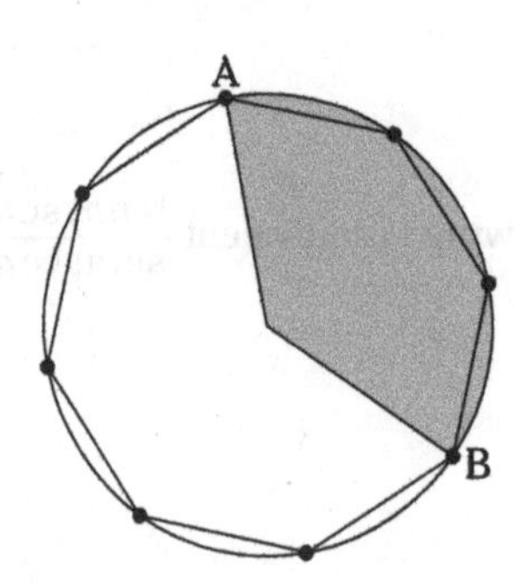

(F) 8π

(G) 16π

(H) 24π

(J) 36π

(K) 64π

DO NOT TURN THE PAGE UNTIL TOLD TO DO SO **STOP** DO NOT RETURN TO A PREVIOUS TEST

Chapter 17
Practice Test 2: Answers and Explanations

You've likely come to this chapter because you just completed Practice Test 2 in Chapter 16. I hope you fared well. To find out, spend some time with this chapter to check your answers. I provide detailed explanations along with a bare-bones answer key. The answer key is great for when you're checking your answers in a hurry, but I suggest that you review the detailed explanations for each question — even those you answered correctly. The explanations show you how to calculate each problem and often provide tips and tricks.

Mathematics Test

1. **D.** Two candy bars and two bags of potato chips cost $4.00, and when you add in one more candy bar, the cost goes up by $0.75. So you know that a candy bar costs $0.75. Thus, two candy bars cost $1.50. Deduct this amount from $4.00 to determine how much two bags of potato chips cost: $\$4.00 - \$1.50 = \$2.50$. So one bag of potato chips costs $1.25.

2. **G.** Start with the highest number in the answers and work your way down: The number 10 is a factor of 60 but not 64, so Choice (K) is wrong. The number 8 isn't a factor of 60, so Choice (J) is wrong. The number 6 is a factor of 60 but not 64, so Choice (H) is wrong. The number 4 is a factor of both 60 and 64, so the correct answer is Choice (G).

3. **D.** The 5 children each received 24 stones, so the total number was $5 \times 24 = 120$. If 6 children had been present, each would have received $120 \div 6 = 20$.

4. **H.** The perimeter of the square is 20, so each side is 5 (because $20 \div 4 = 5$). Use the area formula to find the area of the square:

   $$A = s^2 = 5^2 = 25$$

   The shaded region is half the area of the square: $25 \div 2 = 12.5$.

5. **D.** Exactly 9 people ordered beef stroganoff (25% of $36 = 9$), and 17 people ordered chicken divan. So $9 + 17 = 26$ people *didn't* order linguini primavera. As a result, you know that $36 - 26 = 10$ people ordered this pasta dish.

6. **K.** Begin by substituting -1 for v and 4 for w:

$$4v\left(w^2 - 3vw\right) = 4(-1)\left((4)^2 - 3(-1)(4)\right)$$

Evaluate using the standard order of operations:

$$= 4(-1)(16 + 12) = 4(-1)(28) = -112$$

7. **A.** If you let w equal the amount of time it took Maeve to walk across the width of the field, then $4w$ equals the time she took to walk the length. Maeve walked around the perimeter of the field in 10 minutes, so plug these values into the equation for the perimeter of a rectangle:

$$P = 2l + 2w$$
$$10 \text{ minutes} = 2(4w) + 2w$$

Simplify the right side of the equation:

$$10 \text{ minutes} = 8w + 2w$$
$$10 \text{ minutes} = 10w$$

Divide both sides by 10:

$$1 \text{ minutes} = w$$

So Maeve would take 1 minute to walk across the width of the field.

8. **G.** From left to right, count *down* 6 and *over* 3. Then change this statement into a fraction: $\frac{-6}{3}$. (*Down* means you need a negative and *over* indicates the division bar.)

Now reduce the fraction to get your answer:

$$-\frac{6}{3} = -2$$

9. **D.** Isolate the absolute value on one side of the equation:

$$|2v - 13| + 6 = 9$$
$$|2v - 13| = 3$$

Now split the absolute value into two separate equations and solve both:

$$2v - 13 = 3 \qquad 2v - 13 = -3$$
$$2v = 16 \qquad 2v = 10$$
$$v = 8 \qquad v = 5$$

The value of $8 + 5 = 13$.

10. H. Let x be Damien's score on the fourth day. So his four scores were 93, 92, 89, and x; his average was 90. Plug these numbers into the formula for the mean and simplify:

$$\text{Mean} = \frac{\text{Sum of values}}{\text{Number of values}}$$
$$90 = \frac{93 + 92 + 89 + x}{4}$$
$$90 = \frac{274 + x}{4}$$

Solve for x:

$$360 = 274 + x$$
$$86 = x$$

11. B. Multiply both sides of the equation by 2 to eliminate the fraction, and then simplify:

$$7 - 3p \geq \frac{5}{2}$$
$$14 - 6p \geq 5$$
$$-6p \geq -9$$

Divide both sides by –6 and reverse the inequality, and then reduce the resulting fraction:

$$p \leq \frac{-9}{-6}$$
$$p \leq \frac{3}{2}$$

12. K. Line M and line N are parallel, so the following equivalencies are true:

$$\angle a = \angle e$$
$$\angle c = \angle f$$

Thus, $\angle a + \angle b + \angle f = \angle a + \angle b + \angle c = 180°$. So you know that Choice (F) is wrong. And $\angle a + \angle d = \angle e + \angle d = 180°$, so Choices (G) and (J) are both wrong as well. The three angles in a triangle add up to 180°, so $\angle b + \angle e + \angle f = 180°$, thus Choice (H) is wrong. By the process of elimination, the correct answer is Choice (K).

13. A. Plug the coordinates for the two points into the distance formula:

$$\text{Distance} = \sqrt{(x_2 - x_1)^2 + (y_2 - y_1)^2} = \sqrt{(6 - (-3))^2 + (-5 - 7)^2}$$

Then simplify to get your answer:

$$= \sqrt{9^2 + (-12)^2} = \sqrt{81 + 144} = \sqrt{225} = 15$$

14. F. To begin, find the greatest common factors (GCF) of the coefficient, the x, and the y for the expression $4x^2y^4 - 12x^3y^2 - 8xy^3$:

The GCF of 4, 12, and 8 is 4.

The GCF of x^2, x^3, and x is x.

The GCF of y^4, y^2, and y^3 is y^2.

Thus, $2x^2$ isn't a factor of $4xy^2$.

15. D. Of the 126 students who applied for the scholarship, 9 received it and 117 didn't. You can set up and simplify the ratio like this:

$$\frac{9}{117} = \frac{1}{13}$$

16. K. A parallelogram is a quadrilateral, so its angles add up to 360°. Opposite angles are equal, so any two adjacent angles add up to 180°. Thus, you can set up the following equation:

$$\begin{aligned} z - 15 + 2z &= 180 \\ 3z - 15 &= 180 \\ 3z &= 195 \\ z &= 65 \end{aligned}$$

Therefore

$$\angle D = z° - 15° = 65° - 15° = 50°$$

Because opposite angles in a parallelogram are equal, $\angle D = \angle B$, so $y = 50$.

17. E. To begin, solve $2x - y = 32$ for y in terms of x:

$$\begin{aligned} 2x - y &= 32 \\ -y &= -2x + 32 \\ y &= 2x - 32 \end{aligned}$$

Next, substitute $2x - 32$ for y in $5x + 3y = 14$ and solve for x:

$$\begin{aligned} 5x + 3(2x - 32) &= 14 \\ 5x + 6x - 96 &= 14 \\ 11x &= 110 \\ x &= 10 \end{aligned}$$

Now substitute 10 for x in $y = 2x - 32$:

$$y = 2(10) - 32 = 20 - 32 = -12$$

Therefore, $xy = (10)(-12) = -120$.

18. K. Begin by getting a common denominator of $3b$ on the left side of the equation, and then add the two fractions and simplify:

$$\begin{aligned} \frac{a}{b} + \frac{a+2}{3b} &= \frac{1}{4} \\ \frac{3a}{3b} + \frac{a+2}{3b} &= \frac{1}{4} \\ \frac{3a + a + 2}{3b} &= \frac{1}{4} \\ \frac{4a + 2}{3b} &= \frac{1}{4} \end{aligned}$$

Now cross-multiply to remove the fractions:

$$4(4a + 2) = 3b(1)$$

Simplify and isolate the a term:

$$\begin{aligned} 16a + 8 &= 3b \\ 16a &= 3b - 8 \end{aligned}$$

Finally, divide both sides by 16 to get your answer:

$$a = \frac{3b - 8}{16}$$

19. **A.** Translate the statement "*p* percent of 250 is 75" into an equation and solve for *p*:

$$p(0.01)(250) = 75$$
$$2.5p = 75$$
$$p = 30$$

Thus, 75% of 30 = 22.5.

20. **F.** To find the slope of $9x - 3y = 10$, put the equation in the slope-intercept form:

$$9x - 3y = 10$$
$$-3y = -9x + 10$$
$$y = 3x - \frac{10}{3}$$

Therefore, the slope of the line is 3.

21. **D.** To find the coordinates, simply use the midpoint formula:

$$\text{Midpoint} = \left(\frac{x_2 - x_1}{2}, \frac{y_2 - y_1}{2}\right) = \left(\frac{-6 + 9}{2}, \frac{1 + 8}{2}\right) = \left(\frac{3}{2}, \frac{9}{2}\right)$$

22. **J.** To begin, split $|6 - 4n| > 1$ into two equations and remove the absolute value bars:

$$6 - 4n > 1 \qquad 6 - 4n < -1$$

Solve the first equation:

$$6 - 4n > 1$$
$$-4n > -5$$
$$n < \frac{5}{4}$$

Then solve the second equation:

$$6 - 4n < 1$$
$$-4n < -7$$
$$n > \frac{7}{4}$$

Therefore, $n < \frac{5}{4}$ or $n > \frac{7}{4}$.

23. **E.** The sine equals the opposite side over the hypotenuse:

$$\sin\theta = \frac{O}{H} = \frac{2}{\sqrt{29}}$$

To remove the radical from the denominator, multiply both the numerator and the denominator by $\sqrt{29}$:

$$= \frac{2}{\sqrt{29}} \cdot \frac{\sqrt{29}}{\sqrt{29}} = \frac{2\sqrt{29}}{\sqrt{29}}$$

24. **G.** First, use the formula for the surface area to find the radius:

$$
\begin{aligned}
A &= 4\pi r^2 \\
36\pi &= 4\pi r^2 \\
36 &= 4r^2 \\
9 &= r^2 \\
3 &= r
\end{aligned}
$$

Next, plug the radius into the formula for the volume:

$$V = \frac{4}{3}\pi r^3 = \frac{4}{3}\pi(3)^3 = \frac{4}{3}\pi(27) = 36\pi$$

25. **A.** First, isolate the radical on the left side:

$$
\begin{aligned}
\sqrt{2x+3} - 1 &= x \\
\sqrt{2x+3} &= x + 1
\end{aligned}
$$

Next, square both sides of the equation (be sure to square the whole side in each case):

$$
\begin{aligned}
(\sqrt{2x+3})^2 &= (x+1)^2 \\
2x + 3 &= x^2 + 2x + 1
\end{aligned}
$$

Now simplify and solve for x:

$$
\begin{aligned}
3 &= x^2 + 1 \\
2 &= x^2 \\
\pm\sqrt{2} &= x
\end{aligned}
$$

26. **G.** Begin by listing the factors of 75:

Factors of 75:	1	3	5	15	25	75

Thus, m and n could be any of these six values. A bit of trial and error working with these numbers reveals that $3 + 15 = 18$.

27. **B.** The four percentages shown in the graph add up to $23\% + 21\% + 22\% + 25\% = 91\%$. Thus, Zach owns the remaining 9%. When he divides this percent into four equal parts, each part will be $9\% \div 4 = 2.25\%$. When this amount is added to Bell's, her total shares will be $25\% + 2.25\% = 27.25\%$.

28. **K.** According to the chart, the first balloon costs $4.00, so begin building your function with $f(x) = 4$. Each balloon after the first costs an additional $0.50, so you need to subtract 1 from x (to account for the first balloon) and then multiply this by 0.50:

$$f(x) = 4 + 0.50(x - 1)$$

Simplify the function to get the correct answer:

$$
\begin{aligned}
f(x) &= 4 + 0.50x - 0.50 \\
f(x) &= 3.50 + 0.50x \\
f(x) &= 0.5x + 3.5
\end{aligned}
$$

29. A. Solve $4x^2 - 3x - 1 = 0$ by factoring:

$$4x^2 - 3x - 1 = 0$$
$$(4x + 1)(x - 1) = 0$$

Split this into two equations and solve for x:

$$4x + 1 = 0 \qquad x - 1 = 0$$
$$4x = -1 \qquad x = 1$$
$$x = -0.25$$

Therefore, the sum of these two values is $-0.25 + 1 = 0.75$.

30. H. First, use matrix addition:

$$[4 \ \ 5] + [7 \ \ -3] = [11 \ \ 2]$$

Now multiply the result by 2:

$$2[11 \ \ 2] = [22 \ \ 4]$$

31. D. The equation for the line is $y = \frac{1}{3}x + 3$, so the y-intercept is 3; therefore, $k = 3$. To find the value of n, plug in n for x and 5 for y in the equation:

$$y = \frac{1}{3}x + 3$$
$$5 = \frac{1}{3}n + 3$$
$$2 = \frac{1}{3}n$$
$$6 = n$$

Therefore, $n = 6$, so $kn = 3 \times 6 = 18$.

32. F. Let a equal Alfred's number and r Rani's. Then, translate the statements in the question into two equations as follows:

$$5a + 2r = 300$$
$$2a + 3r = 252$$

Multiply the first equation by 3 and the second equation by 2, and then subtract the second from the first:

$$15a + 6r = 900$$
$$-\ 4a + 6r = 504$$
$$\overline{\quad 11a = 396}$$

Solve for a:

$$a = 36$$

Plug in 36 for a in either equation (the second equation looks easier), and then solve for r:

$$2(36) + 3r = 252$$
$$72 + 3r = 252$$
$$3r = 180$$
$$r = 60$$

Therefore, the sum of the two numbers is $36 + 60 = 96$.

33. **C.** To begin, put $2x - 4y = 13$ into the slope-intercept form:

$$\begin{aligned} 2x - 4y &= 13 \\ -4y &= -2x + 13 \\ y &= \frac{1}{2}x - \frac{13}{4} \end{aligned}$$

The slope of this line is $\frac{1}{2}$. Thus, the slope of the line you're looking for is the negative reciprocal of $\frac{1}{2}$, which is –2. This line passes through the origin, so its y-intercept is 0. Plug these two values into the slope-intercept form:

$$\begin{aligned} y &= mx + b \\ y &= -2x + 0 \\ y &= -2x \end{aligned}$$

34. **J.** Let x equal the score that Gerry received on each of his first two exams. Plug in his scores, the number of exams he took, and his average into the formula for the mean:

$$\begin{aligned} \text{Mean} &= \frac{\text{Sum of values}}{\text{Number of values}} \\ 93 &= \frac{2x + 94 + 85 + 90}{5} \end{aligned}$$

Solve for x by first multiplying both sides of the equation by 5 to eliminate the fraction:

$$\begin{aligned} 465 &= 2x + 269 \\ 196 &= 2x \\ 98 &= x \end{aligned}$$

35. **A.** To begin, multiply both sides by $y + 5$ to eliminate the fraction:

$$\begin{aligned} \frac{2x + 3y - 19}{y + 5} &= 3 \\ 2x + 3y - 19 &= 3(y + 5) \end{aligned}$$

Distribute on the right side:

$$2x + 3y - 19 = 3y + 15$$

Now subtract $3y$ from both sides, and then solve for x:

$$\begin{aligned} 2x - 19 &= 15 \\ 2x &= 34 \\ x &= 17 \end{aligned}$$

36. **H.** From the first experiment to the second:

The value of w is multiplied by 3.

The value of x is divided by 2.

The value of y is multiplied by 2.

The value of z is multiplied by 3.

Thus, the only two possible conclusions would be that w is directly proportional to z and that x is inversely proportional to y.

37. **A.** The *amplitude* is the measure of a wave from the vertical center to the crest. Therefore, this wave has an amplitude of 0.5.

38. **K.** $\overline{OD}$ and $\overline{OE}$ are both radii of the circle, so they're the same length. $\overline{AB}$ is tangent at *D*, so $\overline{AB}$ and $\overline{OD}$ are perpendicular to each other. Similarly, $\overline{BC}$ is tangent at *E*, so $\overline{BC}$ and $\overline{OE}$ are perpendicular to each other. Thus, three angles of *DBEO* are right angles, so the fourth angle is also a right angle. And adjacent sides $\overline{OD}$ and $\overline{OE}$ are the same length, so *DBEO* is a square.

39. **E.** The transformation to move $f(x)$ three units to the right is $f(x-3)$. Then, to reflect this across the *x*-axis, change it to $-f(x-3)$.

40. **G.** The parabola is concave down, so a is negative. It's shifted to the right, so a and b have different signs; therefore, b is positive. It crosses the *y*-axis above the origin, so c is positive. Therefore, a can't be greater than b.

41. **D.** The value inside the radical must be greater than or equal to 0. Thus, if $x = 0$, the value inside the radical is:

$$x^2 - 9 = 0^2 - 9 = -9$$

This value is impossible, so you can rule out Choices (A), (B), and (C). Additionally, the value of the denominator can't equal 0. If $x = 3$, the value of the denominator is

$$\sqrt{x^2 - 9} = \sqrt{3^2 - 9} = \sqrt{9 - 9} = 0$$

This value is also impossible, so Choice (E) is ruled out as well, leaving Choice (D) as the correct answer.

42. **J.** The sequence includes all multiples of 3, including 57 and 72, so Choices (H) and (K) are ruled out. The sequence also includes every number that's 1 added to a multiple of 3, which includes $34\,(33+1)$ and $43\,(42+1)$, so Choices (F) and (G) are ruled out. By the process of elimination, Choice (J) is the correct answer.

43. **A.** Ninety numbers exist in the range from 10 to 99, and 18 of them are divisible by 5. Place these two numbers into the formula for probability:

$$\text{Probability} = \frac{\text{Target outcomes}}{\text{Total outcomes}} = \frac{18}{90} = \frac{1}{5}$$

44. **F.** The base of the triangle is the distance from $(-5, -3)$ to $(-2, -3)$, so it equals 3 units. The height of the triangle extends from $y = -3$ to $y = 4$, so it measures out to be 7 units. Plug these values into the formula for the area of a triangle:

$$A = \frac{1}{2}bh = \frac{1}{2}(3)(7) = 10.5$$

45. **B.** Every triangle has three angles that add up to 180°, and an isosceles triangle has two equivalent angles. Therefore, the three angles of this triangle are either $40° + 40° + 100°$ or $40° + 70° + 70°$. In any case, this triangle can't include a 50° angle.

46. **H.** To begin, FOIL $5 + 6i$ and $3 - 2i$, and then combine like terms:

$$(5+6i)(3-2i) = 15 - 10i + 18i - 12i^2 = 15 + 8i - 12i^2$$

Now substitute -1 for i^2 and combine like terms again:

$$= 15 + 8i - 12(-1) = 15 + 8i + 12 = 27 + 8i$$

47. C. $\overline{KL} \perp \overline{MO}$, and $\overline{KL}$ is a chord of the circle, so $\overline{MO}$ bisects $\overline{KL}$ and $\overline{KM} = 6$ and $\overline{ML} = 6$. Thus, you can make a right triangle as follows:

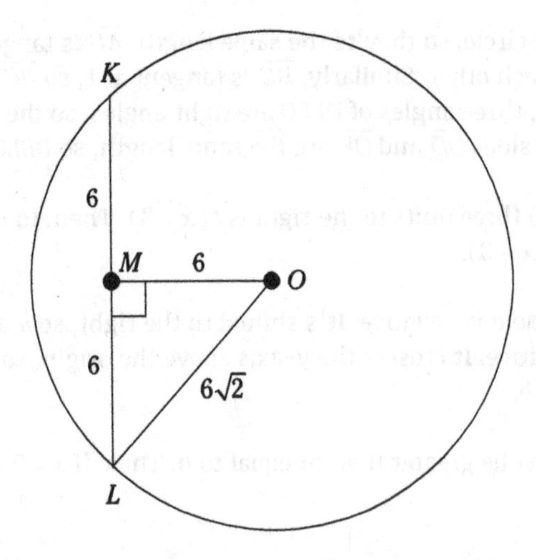

This right triangle has two sides with the length of 6, so its hypotenuse is $6\sqrt{2}$. This value is also the radius of the circle, so plug it into the formula for the area of a circle:

$$A = \pi r^2 = \pi r^2 = \pi\left(6\sqrt{2}\right)^2 = 72\pi$$

48. J. Eldridge opened her account with $1,000, so the line on the graph begins above the origin, which rules out Choices (G) and (H). The withdrawal caused the amount in the account to decrease, so the slope of the line decreases during this month. So Choice (F) is wrong. And in every month except the month with the withdrawal, the upward slope of the line is consistent, which rules out Choice (K). Through the process of elimination, you know that Choice (J) is the correct answer.

49. C. Eldridge started with $1,000, made 8 deposits of $200 each $(8 \times 200 = 1{,}600)$, and withdrew $350, so $1{,}000 + 1{,}600 - 350 = 2{,}250$.

50. H. Let *x* equal the original price for the shoes. The sale price was 0.7*x*, to which a 5% tax was added. So Martha paid 1.05(0.7*x*), and this amount was $58.80. Thus, you set up the following equation:

$$1.05(0.7x) = 58.8$$

Simplify and solve for *x*:

$$0.735x = 58.8$$
$$x = \frac{58.8}{0.735}$$
$$x = 80$$

Therefore, Martha paid $80 for the shoes.

51. E. To begin, use the identity $\tan n = \frac{\sin n}{\cos n}$ to substitute $\frac{\sin n}{\cos n}$ for tan *n*:

$$\frac{\tan n \csc n}{\sin n \sec n} = \frac{\sin n \csc n}{\cos n \sin n \sec n}$$

So you can cancel sin n in the numerator and denominator:

$$= \frac{\csc n}{\cos n \sec n}$$

Now the identity $\cos n = \frac{1}{\sec n}$ means that $\cos n \ \sec n = 1$; therefore, the denominator is 1, so you can drop it:

$$= \frac{\csc n}{1} = \csc n$$

52. **F.** To get a sense of how to proceed, add a few lines to the given figure:

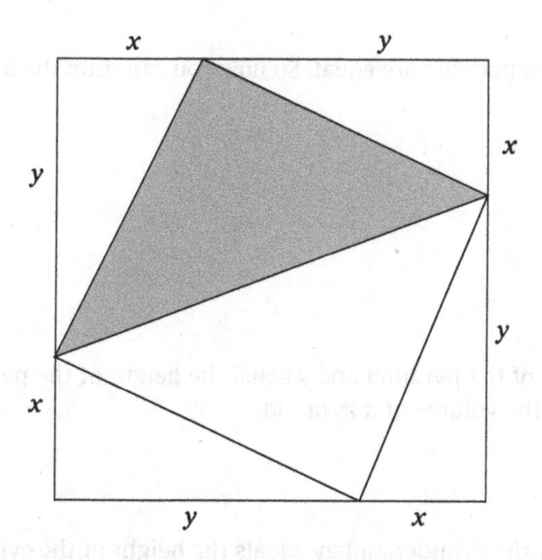

Notice that the shaded triangle is now half of an inner square. Each side of this square is the hypotenuse of a right triangle with legs of length x and y. Use the Pythagorean theorem to find the length of this hypotenuse:

$$a^2 + b^2 = c^2$$
$$x^2 + y^2 = c^2$$
$$\sqrt{x^2 + y^2} = c$$

Thus, the length of a side of this square is $\sqrt{x^2 + y^2}$, so its area is $x^2 + y^2$. The shaded triangle is half of this square, so its area is $\frac{x^2 + y^2}{2}$.

53. **C.** Multiply the two equations together:

$$\frac{f}{g} \cdot \frac{g}{h} = \frac{1}{4} \cdot \frac{2}{5}$$
$$\frac{f}{h} = \frac{2}{20}$$
$$\frac{f}{h} = \frac{1}{10}$$

54. **K.** Begin by squaring both sides of the equation to eliminate the radical. I do this in two steps to keep things clear:

$$49^{3y} = \sqrt{7^{y+1}}$$
$$\left(49^{3y}\right)^2 = \left(\sqrt{7^{y+1}}\right)^2$$
$$49^{6y} = 7^{y+1}$$

Now substitute 7^2 for 49 on the left side. Again, I do this in two steps:

$$\left(7^2\right)^{6y} = 7^{y+1}$$
$$7^{12y} = 7^{y+1}$$

Because the bases are both 7, the exponents are equal. So now you can drop the bases:

$$12y = y + 1$$

Solve for *y*:

$$11y = 1$$
$$y = \frac{1}{11}$$

55. **B.** Let *x* equal the area of the base of the pyramid and *y* equal the height of the pyramid. Plug these values into the formula for the volume of a pyramid:

$$V = \frac{1}{3}A_bh = \frac{1}{3}xy$$

So *x* equals the area of the base of the cylinder and 2*y* equals the height of the cylinder. Now plug these values into the formula for the area of a cylinder, substituting *x* for πr^2:

$$V = \pi r^2 h = x(2y) = 2xy$$

Make a fraction using the area of the pyramid as the numerator and the area of the cylinder as the denominator, and then cancel out *xy* in the numerator and denominator:

$$\frac{\frac{1}{3}xy}{2xy} = \frac{\frac{1}{3}}{2}$$

Multiply the numerator and the denominator by 3 to eliminate the extra fraction:

$$\frac{1}{6}$$

56. **K.** Multiply both sides of the equation by $x + 2$ to remove the fraction, and then place all terms on one side of the equation:

$$x = \frac{2k}{x+2}$$
$$x^2 + 2x = 2k$$
$$x^2 + 2x - 2k = 0$$

The result is a standard quadratic equation, where $a=1$, $b=2$, and $c=-2k$. Use the quadratic formula to solve:

$$x=\frac{-b\pm\sqrt{b^2-4ac}}{2a}=\frac{-2\pm\sqrt{2^2-4(1)(-2k)}}{2(1)}$$

Now simplify:

$$=\frac{-2\pm\sqrt{4+8k}}{2}=\frac{-2\pm2\sqrt{1+2k}}{2}=-1\pm\sqrt{1+2k}$$

Thus, either of the following two solutions is valid:

$$-1+\sqrt{1+2k} \qquad -1-\sqrt{1+2k}$$

So in the first solution, reversing the two terms gives you $\sqrt{1+2k}-1$.

57. **D.** Begin by factoring on both sides of the equation:

$$\frac{a^2+2ab+b^2}{a^2-b^2}=2a+2b$$

$$\frac{(a+b)(a+b)}{(a+b)(a-b)}=2(a+b)$$

Cancel $(a+b)$ in the numerator and denominator:

$$\frac{(a+b)}{(a-b)}=2(a+b)$$

Next, cancel $(a+b)$ on both sides of the equation:

$$\frac{1}{(a-b)}=2$$

Multiply both sides by $(a-b)$:

$$1=2(a-b)$$

Now divide both sides by 2:

$$\frac{1}{2}=a-b$$

58. **J.** The value of a linear function inside absolute value bars can never be less than 0. Thus, a function with a range of $f(x)\geq 4$ takes this minimum value and adds 4 to it. Therefore, $f(x)=|x-4|+4$ can never be less than 4, so the correct answer is Choice (J).

59. **C.** To begin, put the logarithm in exponential form:

$$\log_a\sqrt{b}=\frac{1}{4} \qquad \text{means} \qquad a^{\frac{1}{4}}=\sqrt{b}$$

Square both sides to remove the radical on the right side:

$$\left(a^{\frac{1}{4}}\right)^2=b$$

$$a^{\frac{1}{2}}=b$$

Now square both sides again to isolate *a*:

$$\left(a^{\frac{1}{2}}\right)^2 = b^2$$

$$a = b^2$$

60. **H.** The arc length from *A* to *B* is 6π. The angle from *A* to *B* is $\frac{3}{8}$ of the circle's total of 360°, which equals 135°. Plug these values into the formula for arc length:

$$\text{Arc length} = \text{degrees}\left(\frac{\pi r}{180}\right)$$

$$6\pi = 135\left(\frac{\pi r}{180}\right)$$

Solve for the radius *r*:

$$6\pi = \frac{3\pi r}{4}$$

$$24\pi = 3\pi r$$

$$8 = r$$

Now plug in 8 for *r* in the formula for the area of a circle:

$$\text{Area} = \pi r^2 = \pi(8)^2 = 64\pi$$

The area of the shaded region is $\frac{3}{8}$ of this value:

$$64\pi\left(\frac{3}{8}\right) = 24\pi$$

Answer Key for Practice Test 2

1.	D	21.	D	41.	D
2.	G	22.	J	42.	J
3.	D	23.	E	43.	A
4.	H	24.	G	44.	F
5.	D	25.	A	45.	B
6.	K	26.	G	46.	H
7.	A	27.	B	47.	C
8.	G	28.	K	48.	J
9.	D	29.	A	49.	C
10.	H	30.	H	50.	H
11.	B	31.	D	51.	E
12.	K	32.	F	52.	F
13.	A	33.	C	53.	C
14.	F	34.	J	54.	K
15.	D	35.	A	55.	B
16.	K	36.	H	56.	K
17.	E	37.	A	57.	D
18.	K	38.	K	58.	J
19.	A	39.	E	59.	C
20.	F	40.	G	60.	H

6 The Part of Tens

IN THIS PART . . .

Knowing ten key differences between the ACT and the SAT.

Using a handy checklist to make sure you're set up for success as you head into your ACT.

IN THIS CHAPTER

» **Comparing the upsides and downsides of the ACT and the SAT**

» **Understanding the main features of both tests**

Chapter 18 Ten Key Differences Between the ACT and the SAT

Lots of students take both the ACT and the SAT in order to fulfill the entrance requirements for a wider range of colleges and universities. If you're among this lucky bunch, you're probably studying for both tests even as we speak, so staying clear on the differences between the two tests is important. In this chapter, I provide you with ten important differences between the ACT and the SAT, including advice on how to handle each of them.

Differences in Scoring

The four required sections of the ACT (English, Math, Reading, and Science) are all scored on a scale from 1 to 36. The comprehensive score for all four tests is also scored on the same scale, from 1 to 36. In comparison, the two SAT sections (Reading and Writing, and Math) are both scored on a scale from 200 to 800, with a comprehensive score that's the sum of these two numbers, from 400 to 1600.

Paper vs Online Format

The ACT is still a mostly paper-based test. This means that at the testing center, you'll receive a test booklet plus the familiar bubble-sheet for filling in answers. As of 2024, however, the SAT format is almost exclusively online, requiring the use of a computer.

Each of these formats has its own strengths and weaknesses. If you have a strong preference for an old-fashioned and familiar paper-based test, stick to the ACT. Otherwise, you may want to take the SAT instead.

Keep in mind, though, that the ACT has begun to move online, and that this trend may well accelerate. In short, watch this space for further developments.

Adaptive vs Non-Adaptive Format

Now that the SAT is online, it's become an *adaptive* test. This means that your performance on the first part of each test affects the difficulty level of the questions on the second part.

In contrast, the ACT is still given on paper, and therefore non-adaptive: The questions on your test are simply those found inside your test booklet, so later questions remain unaffected by your performance earlier in the test.

Number of Math Tests

The ACT math test is one 60-minute test that contains 60 questions. The SAT, however, contains two shorter math tests with 22 questions each.

Test Organization

On the ACT, the math section is always the second section out of four. In contrast, on the SAT, the math sections are third and fourth out of four. So, if you like getting your math out of the way early, the ACT is your go-to test.

Existence of Fill-in-the-Blank Questions

The ACT math test contains only multiple-choice questions. Compare this to the SAT, which includes fill-in-the-blank questions. If you're like most students, you probably find multiple-choice questions a bit less intimidating because the choices provide information that can be helpful for getting the right answer.

Answer Multiple-Choice Format

On the ACT, all questions include five possible answers: The odd-numbered questions contain the multiple-choice answers (A) through (E), and the even-numbered questions contain the answers (F) through (K). The answer (I) is omitted (probably because it looks too much like the number 1, which could be confusing in a math test). In contrast, all the multiple-choice questions on the SAT include four possible answers, labeled (A) through (D).

The ACT convention is useful for helping prevent a common nightmare that many students (including myself) have experienced at least once on printed tests: finding that they skipped a question early in the test but forgot to leave space for it on their answer sheets. At this point, they lose precious minutes as they panic, erase, erase, scribble, scribble, and finally continue the test. Not fun.

Because the ACT alternates ABCDE questions with FGHJK questions, you're far less likely to make this mistake. For example, if you skip over an ABCDE question, your next question will be an FGHJK question: Whatever answer you choose, you won't be able to enter it in the space for the question you skipped over.

Amount of Advanced Math

The ACT contains questions that test your understanding of advanced math concepts that aren't tested on the SAT. Three key areas are matrices, logarithms, and imaginary (and complex) numbers. I discuss all these topics in Chapter 12.

The downside of these questions is obvious: If you haven't studied this stuff, you won't know how to answer these questions, so just move on. But here's the upside: Although this material is advanced, if you already know it, the questions tend to be on the easy side. For example, logarithms aren't easy, but if you've studied them in either Algebra 2 or Pre-Calculus, you'll probably find most ACT questions easier than many of your homework problems.

Difficulty of the Reading Test

If you're a strong reader, you may want to take the ACT to boost your score.

The ACT reading test is tough, containing four passages that each require you to answer ten questions. It's also fast, giving you a mere 35 minutes to finish. That's less than one minute per question!

In contrast, the new SAT Reading and Writing test includes no long reading passages. While this feature may sound attractive, if you're confident about your reading skills, the ACT may give you an advantage over your competition.

Presence of the Science Test on the ACT

The ACT includes a Science test, measuring your ability to read and understand scientific material, including tables and graphs. This test also includes a few questions that require you to recall information from your study of science that isn't explicitly included on the test.

In contrast, the SAT Reading and Writing test includes a few short passages related to science, each of which may include one or more tables or graphs.

So, if science is one of your strong suits, you may want to take the ACT. If not, you might prefer to gravitate toward the SAT.

IN THIS CHAPTER

» Going through a checklist of items to do before your ACT

» Preparing yourself physically and mentally

Chapter 19
Ten Items to Check Off before Taking the ACT

You can't be too organized. (Well, my Great Aunt Ida can. I mean, come on, who really alphabetizes socks?) So if you're not already the kind of person who keeps argyles on the left side of the drawer and woolies on the right, here are ten items to check off as you approach the ACT.

I Received My ACT Admission Ticket, and I Put It in a Safe Place

As with movies, but with far less entertainment value, the ACT requires a ticket to get in. You should receive your ticket well in advance of the test after you've registered and paid. When you receive it, put your ticket in a safe, dry place until the night before the test. The last thing you need the day of your test is a frantic search for a vanished ticket.

I'm 100 Percent Sure about the Date and Time of My ACT

When you receive your ticket, check the date and time of your ACT and put it on a calendar you check regularly. Or, if you keep appointments on your phone, set an alarm a few days before to make sure you don't miss the test.

WARNING

If you miss your ACT, you lose the money you spent. You can't receive a refund or postponement after the fact.

I Know How to Get There, Too

If the test center for your ACT happens to be your school, finding it should be a no-brainer. But if it's held someplace else, make sure you're clear on how to get there.

I Purchased the Calculator I Intend to Use

Buy a calculator well in advance of the ACT so you have time to practice with it. At a minimum, this calculator should be able to give you the square root (radical) of a number. I provide a variety of considerations about calculators in Chapter 2.

I Feel Comfortable Using My Calculator

After you've picked out and purchased the calculator you intend to use on the ACT, don't let it just sit there in its hard-to-open plastic shell until the day before the test. Be sure to use it when studying and taking practice tests. All calculator models are slightly different, so using one before you're under the time pressure of a test is usually the best way to find out all its little quirks.

I Have a Backpack Ready to Go by the Door

The night before your ACT, pack a backpack (or some other container to carry stuff in) with everything you need, including the following:

- Admission ticket
- Identification
- Plenty of sharpened #2 pencils
- A calculator that you know how to use (with fresh batteries)
- Extra batteries for your calculator
- Something to eat or drink during your break, which happens just after you finish your math test

I Picked Out My Clothes for the Morning

I know I probably sound like your mom, but she's right on this one: If you lay out your clothes the night before your test, you'll have one less thing to think about the morning of your ACT. No matter what the weather is in the morning, be sure to wear a few layers of clothing in case the temperature of the room is too hot or too cold for your liking. You're going to be stuck in that one room for four or five hours, so you want to be sure you're comfortable.

I'm Having a Relaxing Night before the Test

The night before the ACT is yours to do with as you choose. If you work, get the night off well in advance so you have plenty of time to relax and get a good night's sleep. Then spend this time in whatever way you enjoy — with friends or alone, watching a movie or playing a video game, walking on the beach or biking around your neighborhood, playing with your dog or feeding your fish. Do whatever makes you feel calm and rested.

WARNING

I don't recommend studying the night before. And, most important, don't take a practice test! If you must study, limit the time to half an hour — just enough to review a few formulas or look over some practice problems.

My Alarm Is Set for an Early Rise

Set your alarm early enough to allow plenty of time to get ready, eat breakfast, and get out the door and on the road. Better to arrive early and wait around for a few minutes than arrive late and run in the door panicked, out of breath, and unfocused.

I'm Focusing on My Breathing

If you start to feel nervous while waiting for the ACT to start, take a few nice deep breaths (but not too many — you don't want to hyperventilate). Deep breaths give you oxygen, which is always a good thing. Oxygen helps to move adrenalin — the hormone that accounts for that shaky feeling you get when scared or nervous — out of your bloodstream. Breathe as you sit down to begin the test. Breathe as you begin the test. And along the way, if you notice anxiety beginning to creep in, take some more deep breaths.

Index

D

E

F

G

H

I

L

M

N

O

P

Q

R

S

T

V

W

X

Y

Z

About the Author

Mark Zegarelli is the author of *SAT Math For Dummies* (Wiley), *Basic Math & Pre-Algebra For Dummies* (Wiley), and nine other books in the *For Dummies* series. He holds degrees in both English and math from Rutgers University. Most recently, he provides online lessons for kids as young as four years old, showing them easy ways to understand — and enjoy! — square numbers and square roots, factors, prime numbers, fractions, and even basic algebra concepts.

Dedication

For my good friend David Feaster — as always, with love, laughter, and light.

Author's Acknowledgments

This is my eleventh *For Dummies* book, and my eleventh positive experience working with a first-rate team of editors who make the process of writing so very pleasurable and productive. Thanks so much to my Wiley editors Kezia Endsley, Kristie Pyles, Elizabeth Stillwell, and Liz McKee. Thanks as always to my assistant Chris Mark for helping to keep me on track. More thanks to technical editor Ana Teodorescu for keeping me accurate and finely tuned to the needs of readers who are studying for the ACT. Additional thanks to Tamilmani Varadharaj for the final layout of the book.

And, as usual, many thanks to the good folks at Castro Coffee Company for pouring every much-needed cup with care and devotion.

Publisher's Acknowledgments

Senior Acquisitions Editor: Elizabeth Stillwell

Project Editor: Kezia Endsley

Technical Editor: Ana Teodorescu

Production Editor: Tamilmani Varadharaj

Cover Image: © Ed Bock/Getty Images

PERSONAL ENRICHMENT

9781119187790
USA $26.00
CAN $31.99
UK £19.99

9781119179030
USA $21.99
CAN $25.99
UK £16.99

9781119293354
USA $24.99
CAN $29.99
UK £17.99

9781119293347
USA $22.99
CAN $27.99
UK £16.99

9781119310068
USA $22.99
CAN $27.99
UK £16.99

9781119235606
USA $24.99
CAN $29.99
UK £17.99

9781119251163
USA $24.99
CAN $29.99
UK £17.99

9781119235491
USA $26.99
CAN $31.99
UK £19.99

9781119279952
USA $24.99
CAN $29.99
UK £17.99

9781119283133
USA $24.99
CAN $29.99
UK £17.99

9781119287117
USA $24.99
CAN $29.99
UK £16.99

9781119130246
USA $22.99
CAN $27.99
UK £16.99

PROFESSIONAL DEVELOPMENT

9781119311041
USA $24.99
CAN $29.99
UK £17.99

9781119255796
USA $39.99
CAN $47.99
UK £27.99

9781119293439
USA $26.99
CAN $31.99
UK £19.99

9781119281467
USA $26.99
CAN $31.99
UK £19.99

9781119280651
USA $29.99
CAN $35.99
UK £21.99

9781119251132
USA $24.99
CAN $29.99
UK £17.99

9781119310563
USA $34.00
CAN $41.99
UK £24.99

9781119181705
USA $29.99
CAN $35.99
UK £21.99

9781119263593
USA $26.99
CAN $31.99
UK £19.99

9781119257769
USA $29.99
CAN $35.99
UK £21.99

9781119293477
USA $26.99
CAN $31.99
UK £19.99

9781119265313
USA $24.99
CAN $29.99
UK £17.99

9781119239314
USA $29.99
CAN $35.99
UK £21.99

9781119293323
USA $29.99
CAN $35.99
UK £21.99

dummies.com

dummies®
A Wiley Brand